Duality and Approximation Methods for Cooperative Optimization and Control

Von der Fakultät Konstruktions-, Produktions- und Fahrzeugtechnik
und dem Stuttgart Research Centre for Simulation Technology
der Universität Stuttgart zur Erlangung der Würde eines
Doktor-Ingenieurs (Dr.-Ing.) genehmigte Abhandlung

Vorgelegt von

Mathias Bürger

aus Bad Kissingen

Hauptberichter:	Prof. Dr.-Ing. Frank Allgöwer
Mitberichter:	Prof. Dr. Claudio De Persis
	Prof. Dr.-Ing. Sandra Hirche

Tag der mündlichen Prüfung: 2. Dezember 2013

Institut für Systemtheorie und Regelungstechnik

Universität Stuttgart

2013

Bibliographic information published by the Deutsche Nationalbibliothek

The Deutsche Nationalbibliothek lists this publication in the Deutsche Nationalbibliografie; detailed bibliographic data are available in the Internet at http://dnb.d-nb.de .

ISBN 978-3-8325-3624-4

Logos Verlag Berlin GmbH
Comeniushof, Gubener Str. 47,
10243 Berlin
Tel.: +49 (0)30 42 85 10 90
Fax: +49 (0)30 42 85 10 92
INTERNET: http://www.logos-verlag.de

Für Inga

Acknowledgements

This thesis presents the results of my research at the Institute for Systems Theory and Automatic Control (IST), University of Stuttgart. During my time at the IST, I was fortunate to find a number of people who became teachers, mentors, and friends to me.

First of all, I want to thank Prof. Frank Allgöwer. He was a challenging, but always motivating and supportive advisor and he led me well on my way through the academic world. His continuous advice on a professional, but also on the personal level, was invaluable to me. I also thank him for giving me the freedom to pursue my interests and for creating an open minded and international environment, where research ideas can arise and grow.

I also express my gratitude to Prof. Claudio De Persis (University of Groningen), Prof. Sandra Hirche (TU Munich), and Prof. Arnold Kistner (University of Stuttgart) for their interest in my work, for their encouraging comments, and for being members of my doctoral examination committee. In particular, I sincerely thank Prof. Claudio De Persis for the insightful and constructive discussions we had, our productive collaboration, and all his personal advice.

Prof. Daniel Zelazo (Israel Institute of Technology) had a great influence on my work. I thank him for the great collaboration and the countless hours we spent together in front of the blackboard. His positive attitude was a constant source of motivation for me.

Prof. Giuseppe Notarstefano (University of Lecce) opened me the door into distributed optimization and continuously challenged me with new research questions. Throughout my doctoral studies he has been a great teacher and friend. Without him, my thesis would not have become what it is now.

I also thank Prof. Francesco Bullo (University of California at Santa Barbara) for inviting me to spend three month with his group and for his continuous support before and after my visit. Moreover, I thank Prof. Martin Guay (Queen's University) for encouraging me to enter the research world and for continuously accompanying my scientific career.

I also thank all my colleagues and friends at the IST. I am thankful that I could be part of this very special group of people. Special thanks goes to my long time office mate Matthias Müller, as well as to Florian Bayer, Rainer Blind, Matthias Lorenzen, Max Montenbruck, Marcus Reble, Georg Seyboth and Gerd Schmidt for the excellent cooperation.

I also thank my parents and my sisters for the constant support they gave me during the past years. Finally, I thank Inga for being with me on this journey.

Stuttgart, December 2013
Mathias Bürger

Table of Contents

Abstract

Almost all distributed systems are faced with the challenges of *cooperative optimization and control*. The objective of this thesis is to investigate the role of *duality* and the use of *approximation* methods in cooperative optimization and control.

In this thesis, a general algorithm for convex optimization in networks with asynchronous communication is presented. The main algorithmic idea pursued here is the one of polyhedral approximation, where a complex decision problem is replaced by a sequence of simpler (linear) problems. A distributed algorithm for convex optimization problems is proposed based on this conceptual idea. The algorithm is presented first in an abstract formulation, and is then specified to a family of optimization algorithms, solving for example semi-definite or robust optimization problems.

Following this, the relevance of approximation methods in cooperative predictive control is shown. It is shown that well-known trajectory exchange methods can be understood as polyhedral approximation methods, if the dual of the predictive control problem is considered. With this observation, a novel trajectory exchange method for distributed model predictive control is proposed, based on the previously developed distributed optimization algorithm. This result highlights the importance of approximation methods and duality in cooperative decision making and control.

Duality is also shown to be a concept of major importance for cooperative control in general. The relevance of duality in cooperative control is illustrated on networks of passive systems that are designed to reach output agreement. An "inverse optimality" result is presented, that relates the asymptotic behavior of the dynamic variables to a family of network optimization problems. We identify several pairs of dynamic variables that approach the solutions of dual pairs of network optimization problems and provide in this way a complete duality theory for passivity-based cooperative control.

This duality result contributes towards a general theory for cooperative control and opens the way for understanding complex dynamic phenomena. Building upon the duality result, the dynamic phenomenon of "clustering" is studied. Clustering refers to a situation where only groups of systems (rather than all systems) reach output agreement. A variation of the passivity-based cooperative control framework is considered that leads to such a clustering behavior. The clustering phenomenon is exactly analyzed and predicted, using the proposed duality results. In fact, it is shown that only taking a combined primal and dual perspective enables one to fully understand the clustering phenomenon. In this way, this thesis presents theoretical insights and algorithmic approaches for cooperative optimization and control, and emphasizes the role of convexity and duality.

Deutsche Kurzfassung

Motivation und Zielsetzung

Betrachtet man die technischen Entwicklungen der vergangenen Jahre, kann man sehr schnell eine klare Entwicklungsrichtung erkennen. Technische Systeme bestehen in zunehmendem Maße aus vielen verteilten Komponenten, die flexibel aber zuverlässig miteinander interagieren müssen. Beispiele hierfür lassen sich recht einfach finden. Kooperierende mobile Roboter werden heute beispielsweise zur Warenverteilung in Logistikzentren genutzt und müssen sich dort selbstständig koordinieren (Behnke, 2013). Netzwerke aus sehr vielen, örtlich verteilten Sensoren werden heute genutzt, um Umweltbeeinflussungen oder Verkehrsströme zu beobachten. Die gemessenen Daten müssen oft schon im Netzwerk verarbeitet werden, was eine Kooperation zwischen „intelligenten" Sensoren voraussetzt (Stojmenovic, 2005). In „intelligenten" Stromnetzen sind Generatoren und Verbraucher mit Technik ausgestattet, die es erlaubt, dass das gesamte Netz kooperiert und flexibel auf Veränderungen in den Anforderungen reagiert (Zamora and Srivastava, 2010). Dies sind nur wenige der vielen Beispiele für technische Systeme, bei denen die Interaktion und Kooperation vieler verteilter Einheiten entscheidend für die Funktionalität des gesamten Netzwerks ist. So unterschiedlich die Herausforderungen der genannten Beispiele aus technischer Sicht auch sind - aus methodischer Sicht stellen sie sehr ähnliche Anforderungen. Viele der methodischen Herausforderungen liegen im Gebiet der kooperativen Optimierung und Regelung. *Optimierung* meint das Auswählen der „besten" von mehreren Alternativen, während *Regelung* zum Ziel hat, ein dynamisches System so zu beeinflussen, dass es ein gewünschtes Verhalten zeigt. *Kooperation* zwischen den Teilsystemen ist notwendig, um ein technisches System in einen „global" optimalen Zustand zu bringen. Ein theoretisches Verständnis sowie algorithmische Methoden für eine kooperative Optimierung und Regelung sind eine Grundvoraussetzung für viele moderne technische Systeme. Unter den vielfältigen theoretischen Problemstellungen, die verteilte Systeme bieten, ragen zwei Probleme heraus, die auch von zentraler Bedeutung für diese Arbeit sind.

Eine Herausforderung, der man in beinahe allen verteilten Systemen begegnet, ist, ein endlichdimensionales Optimierungsproblem verteilt zu lösen, d.h. in einem Netzwerk von Prozessoren zu berechnen. Sehr viele Entscheidungsprobleme können direkt als Optimierungsprobleme formuliert werden. Das gilt zum Beispiel für Schätzprobleme in Sensornetzwerken, Ressourcenverteilungsprobleme oder auch prädiktive Optimalsteuerungsprobleme. Die Herausforderung ist für all diese Probleme ähnlich. Die verteilten Einheiten, oder Prozessoren, müssen kooperativ eine optimale Lösung finden, wobei kein einzelner Prozessor die gesamten Informationen besitzt. Eine Lösung kann nur durch einen Informationsaustausch über ein eventuell unzuverlässiges Kommunikationsnetz erreicht werden. Diese allgemeine mathematische Problembeschreibung ist für viele technische Probleme von zentraler Bedeutung. Konsequenterweise hat das Problem der verteilten Optimierung über die vergangenen

Jahre hinweg viel Beachtung gefunden, so dass heute verschiedene Lösungsansätze existieren, siehe z.B. Necoara et al. (2011). Dennoch sind noch viele grundlegenden Fragen offen. Die großen Herausforderungen bleiben zum einen zu klassifizieren, welche Klassen von Optimierungsproblemen effektiv verteilt gelöst werden können, und zum anderen, wie mit einer asynchronen Kommunikation umgegangen werden muss.

Eine zweite große Herausforderung ist, das Verhalten dynamischer Netzwerke zu analysieren oder, aus einer anderen Perspektive, dynamische Netzwerke optimal zu entwerfen. Dynamische Netzwerke können entweder durch physikalische Kopplungen verteilter Systeme entstehen, wie z.B. bei Energienetzwerken, oder aber dadurch, dass eine Regelungsaufgabe eine Kooperation erfordert, wie z.B. beim Kolonnenfahren autonomer Fahrzeuge. Solche Netzwerke dynamischer Systeme werfen vielfältige neue Fragen auf. Sie können beispielsweise ein sehr komplexes dynamisches Verhalten zeigen, selbst wenn die einzelnen Teilsysteme sehr „einfach“ strukturiert sind. Das große Ziel aller Forschungsansätze ist zu verstehen, wie das Zusammenspiel von lokalen Dynamiken, Interaktionsprotokollen und Netzwerk-Topologie das Verhalten des gesamten Netzwerks beeinflusst, siehe Mesbahi and Egerstedt (2010). Um dieses weitgefasste Ziel irgendwann erreichen zu können, ist es zunächst notwendig zu verstehen, unter welchen Bedingungen sich dynamische Netzwerke vorhersagbar und eventuell auch optimal verhalten.

In dieser Arbeit werden verschiedene Problemstellungen in der kooperativen Optimierung und Regelung betrachtet. Aus methodischer Sicht ist ein wichtiges Konzept in dieser Arbeit die *Approximation*. Natürlicherweise ist es oft schwierig, Optimierungsprobleme verteilt zu lösen. Wir verfolgen daher hier die Idee, nicht das Originalproblem direkt anzugehen, sondern stattdessen eine Reihe von Approximationen des Problems zu betrachten. Wir ersetzen das komplexe ursprüngliche Problem durch eine Approximation, die eine klare Struktur hat und somit einfacher verteilt gelöst werden kann. Basierend auf der so berechneten (approximativen) Lösung, wird eine neue, genauere Approximation des Originalproblems erstellt. So führt ein wiederholtes Lösen der einfacheren Approximationen zur Lösung des ursprünglichen Optimierungsproblems. Diese einfache konzeptionelle Idee wird in dieser Arbeit genutzt, um verschiedene Problemstellungen, von robusten Optimierungsproblemen bis hin zu kooperativen prädiktiven Regelungsproblemen, in asynchronen Prozessornetzwerken zu lösen.

Das zweite wichtige Konzept dieser Arbeit ist die *Dualität*. Dualität ist die mathematische Formalisierung der Idee, dass Objekte oder Probleme stets auf zwei verschiedene Arten betrachtet werden können.[1] Die Betrachtung eines Problems von beiden Seiten, der primären (orginalen) und der dualen, ermöglicht es häufig, ein tiefgehendes Verständnis über die Vorgänge und Zusammenhänge zu gewinnen. Wir nutzen diese grundlegende Idee, stets auch das duale Problem zu betrachten, intensiv an verschiedenen Stellen dieser Arbeit. So ist Dualität der Schlüssel, um die entwickelten Optimierungsalgorithmen auf eine neue, größere Klasse von Problemen anzuwenden. Darüber hinaus wird in dieser Arbeit gezeigt, dass Dualitätsrelationen nicht nur in der Optimierungstheorie sondern auch in der kooperativen

[1] So ist zum Beispiel das Problem den kleinsten Abstand eines Punktes zu einer Geraden zu finden, dual zu dem Problem den größten Abstand des Punktes zu Ebenen durch die Gerade zu finden. Löst man eines der beiden Probleme, löst man implizit auch immer das zweite.

Regelung bedeutend sind. Wir charakterisieren Dualitätsrelationen zwischen verschiedenen Variablen in kooperativen Regelungssystemen, indem wir diese Variablen mit Paaren dualer Optimierungsprobleme verknüpfen. Neben der theoretischen Eleganz haben diese Ergebnisse auch eine große praktische Bedeutung, die ausführlich diskutiert wird.

Forschungsbeiträge und Gliederung

In dieser Arbeit werden verschiedene Beiträge für eine Theorie der kooperativen Optimierung und Regelung vorgestellt. Im Folgenden werden zunächst die Hauptbeiträge der Arbeit zusammengefasst. Danach werden die Gliederung der Arbeit und die Beiträge der einzelnen Kapitel im Detail vorgestellt.

Hauptbeiträge

Ein erster Beitrag der Arbeit ist ein neuer Ansatz zur verteilten Optimierung in Prozessornetzwerken mit asynchroner Kommunikation. Wir betrachten hierfür eine allgemeine Klasse verteilter Optimierungsprobleme, in der jedem Prozessor eine konvexe Lösungsmenge zugeordnet wird und in der die Prozessoren gemeinsam eine optimale Lösung in der Schnittmenge der lokalen Mengen berechnen müssen. Wir zeigen an verschiedenen Beispielen, dass diese allgemeine Problemklasse sehr viele konkrete Problemstellungen umfasst, von der Positionsschätzung in Sensornetzwerken bis hin zum optimalen Management von Energienetzen. In dieser Arbeit wird ein konzeptioneller Algorithmus entwickelt, der, unabhängig von der spezifischen Form des Optimierungsproblems, alle Probleme in der allgemeinen Problemklasse in einem asynchronen Kommunikationsnetzwerk lösen kann. Speziell schlagen wir einen Algorithmus vor, bei dem die Prozessoren sequentiell Approximationen der Lösungsmenge berechnen und diese untereinander austauschen. Der vorgeschlagene Algorithmus stellt sehr geringe Anforderungen an das Kommunikationsnetz und an die Prozessoren. Wir zeigen formal die Korrektheit des Algorithmus und seine Robustheit gegen Prozessorausfälle oder Kommunikationsstörungen. Der allgemeine Algorithmus wird danach für verschiedene Repräsentationen des Optimierungsproblems spezifiziert. Es wird gezeigt, wie der allgemeine Algorithmus angepasst werden kann, um beispielsweise semi-definite oder robuste Optimierungsprobleme zu lösen. Mit diesem breiten konzeptionellen Rahmen leistet die vorliegende Arbeit einen Beitrag zu einer grundlegenden Theorie verteilter Optimierung.

Der zweite Beitrag dieser Arbeit bezieht sich auf kooperative prädiktive Regelungsprobleme. Wir betrachten hierfür eine Klasse von konvexen Koordinationsproblemen zwischen dynamischen Systemen. Ein (heuristischer) Standardansatz, solche Probleme verteilt zu handhaben, basiert auf sogenannten Trajektorienaustauschverfahren. Wir präsentieren hier eine neue Interpretation der Trajektorienaustauschverfahren als Approximationsmethoden für das *duale* Optimierungsproblem. Diese duale Interpretation erlaubt es uns zu verstehen, warum Trajektorienaustauschverfahren in der Regel nicht zur optimalen Lösung konvergieren können. Speziell zeigen wir, dass mehr als eine prädizierte Trajektorie pro System gespeichert und ausgetauscht werden muss. Basierend auf diesen Überlegungen stellen wir einen neues Trajektorienaustauschverfahren vor, das auf dem in der Arbeit entwickelten Algorithmus basiert und eine Konvergenz zur optimalen Lösung garantiert.

Diese zwei ersten Beiträge der Arbeit beziehen sich auf das Problem, Optimierungsaufgaben verteilt zu lösen. Die nächsten Beiträge der Arbeit zielen darauf ab, Methoden zur Analyse und für den optimalen Entwurf dynamischer Netzwerke zu entwickeln.

Als dritter Hauptbeitrag der Arbeit wird eine vollständige Dualitätstheorie für passivitätsbasierte kooperative Regelungssysteme entwickelt. Hierfür wird ein allgemeiner Rahmen für kooperative Regelungen eingeführt in dem die Regelungsaufgabe ist, einen Konsens auf den Ausgangsgrößen der Teilsysteme zu erreichen. Wir nutzen eine verfeinerte Version des klassischen systemtheoretischen Konzepts der Passivität, nämlich die Gleichgewichtsunabhängige Passivität (engl. equilibrium independent passivity), um ein kanonisches Modell kooperativer Regelungen zu entwickeln. Das betrachtete Modell ist eine Verallgemeinerung vorhandener Modelle aus der Literatur. Ein Hauptergebnis dieser Arbeit ist die Interpretation der dynamischen Variablen eines solchen kooperativen Regelungssystems in einem allgemeinen netzwerk-theoretischen Sinn. Wir zeigen beispielsweise, dass die Ausgangsgrößen der Teilsysteme als „Potentiale“ verstanden werden können, während die Ausgänge der dynamischen Regler eine Interpretation als „Flüsse“ im Netzwerk zulassen. Wir erreichen diese neue Interpretation der klassischen passivitäts-basierten Regelung, indem wir eine Verbindung zwischen dem asymptotischen Verhalten der dynamischen Systeme und verschiedenen Netzwerkoptimierungsproblemen herstellen. Wir zeigen somit, dass passivitäts-basierte kooperative Regelungen „invers optimal“ sind. Die theoretischen Ergebnisse führen zu einem tiefgehenden Verständnis der Zusammenhänge in kooperativen Regelungssystemen. Darüber hinaus haben sie auch eine klare praktische Relevanz, da sie explizit erklären, wie dynamische Netzwerke entworfen werden müssen, um „optimal“ zu arbeiten.

Als vierten Hauptbeitrag der Arbeit analysieren wir das komplexere dynamische Phänomen des „Clustering“. Clustering bezieht sich auf das Phänomen, dass in einem dynamischen Netzwerk eine Synchronisation der Ausgänge nicht über alle Systeme sondern nur über Gruppen der Systeme erfolgt. Clustering kann sowohl ein gewünschtes Verhalten sein, das durch eine Regelung erzeugt werden soll, als auch ein unerwünschtes Verhalten, das vermieden werden muss. In jedem Fall ist ein grundlegendes Verständnis der Mechanismen, die zu Clustering führen, erstrebenswert. Wir zeigen in dieser Arbeit, dass auch dieses dynamische Phänomen in dem neuen theoretischen Rahmen einfach verstanden werden kann. Zunächst wird gezeigt, dass bereits durch eine kleine Variation der passivitäts-basierten Regelung Clustering auftreten kann. Während Clustering aus dynamischer Sicht ein sehr komplexes Verhalten ist, zeigen wir hier, dass es in dem neuen theoretischen Rahmen der Arbeit sehr einfach verstanden werden kann. Die in der Arbeit entwickelten Dualitätsrelationen sowie die Verbindungen zu Netzwerkoptimierungsproblemen erlauben es uns, das Clustering-Phänomen exakt vorherzusagen. Neben dem grundlegenden Verständnis dynamischer Netzwerke führen diese Ergebnisse somit auch zu konstruktiven Analysemethoden. Wir stellen, basierend auf den Ergebnissen zum dynamischen Clustering, einen neuen optimierungs-basierten Ansatz zur Identifikation hierarchischer Strukturen in Netzwerken vor. Somit trägt die vorliegende Arbeit zu einem grundlegenden Verständnis kooperativer Regelungen bei und stellt neue Methoden zur deren Analyse bereit.

Zusammenfassend liegen die Hauptergebnisse dieser Arbeit in den folgenden vier Bereichen:

(i) Ein Algorithmus zum Lösen von konvexen und robusten Optimierungsproblemen in Prozessornetzwerken mit asynchroner Kommunikation wird entwickelt und analysiert.

(ii) Eine Interpretation von Trajektorienaustauschmethoden in der kooperativen prädiktiven Regelung als duale Version eines Approximationsalgorithmus wird vorgestellt; basierend auf dieser Interpretation, wird ein verbesserter Algorithmus vorgeschlagen.

(iii) Eine vollständige Dualitätstheorie für passivitäts-basierte kooperative Regelungssysteme wird entwickelt.

(iv) Optimierungs-basierte Methoden zur Analyse und Vorhersage von Clustering in dynamischen Netzwerken werden entwickelt.

Gliederung

Die Gliederung der Arbeit orientiert sich an diesen vier Hauptbeiträgen. Im Folgenden werden die Beiträge der einzelnen Kapitel kurz zusammengefasst.

Kapitel 2: Polyhedral Approximation Methods for Cooperative Optimization (Approximationsmethoden zur kooperativen Optimierung)

In diesem Kapitel wird ein konzeptioneller Rahmen zur kooperativen Optimierung in Prozessornetzwerken vorgestellt. Zunächst wird eine allgemeine Klasse verteilter Optimierungsprobleme betrachtet, in denen die Information über das zu lösende Problem nur verteilt in einem Prozessornetzwerk vorliegt. Wir betrachten eine Klasse von Problemen, bei denen jeder Prozessor Kenntnis über eine konvexe Lösungsmenge hat und eine optimale Lösung bezüglich einer linearen Kostenfunktion in der Schnittmenge der Lösungsmengen gefunden werden muss. Für diese allgemeine Problemklasse wird ein neuer Algorithmus, genannt „Cutting-Plane Consensus“ Algorithmus, vorgeschlagen. Um den Algorithmus vorzubereiten, wird zunächst das Problem diskutiert, wie lineare Optimierungsprobleme mit einer eindeutigen Lösung gelöst werden können. Wir stellen zwei Ansätze vor, nämlich die lexikographisch minimale Lösung und die Lösung mit minimaler 2-Norm. Basierend auf dieser Methodik wird dann ein verteilter Optimierungsalgorithmus für allgemeine konvexe Optimierungsprobleme vorgestellt. Der Algorithmus basiert auf einer linearen Approximation der lösbaren Mengen durch sogenannte „Cutting-Planes“ und wird konsequenterweise „cutting-plane consensus “ genannt. Wir zweigen, dass der Algorithmus korrekt funktioniert und eine sehr hohe Robustheit gegen Prozessorausfälle und Störungen in der Kommunikation aufweist. Zunächst wird der Algorithmus in einer allgemeinen, abstrakten Form eingeführt. Danach werden verschiedene Versionen des allgemeinen Algorithmus für spezielle Optimierungsprobleme präsentiert. Es wird beispielsweise gezeigt, wie der Algorithmus semidefinite oder auch robuste Optimierungsprobleme lösen kann. Außerdem wird gezeigt, dass der neue Algorithmus, wenn er zum Lösen linearer Programme genutzt wird, als eine verteilte Version des klassischen Simplex-Algorithmus verstanden werden kann. Dieses Kapitel basiert auf den Veröffentlichungen Bürger et al. (2011b), Bürger et al. (2012b), Bürger et al. (2012a) und Bürger et al. (2014).

Kapitel 3: Dual Cutting-Plane & Trajectory Exchange Optimization (Duale Cutting-Plane Optimierung & Trajektorienaustauschverfahren)

In diesem Kapitel wird die Optimierung in verteilten dynamischen Regelungssystemen mittels Trajektorienaustauschverfahren betrachtet, wie sie häufig in der verteilten prädiktiven Regelung genutzt werden. Wir analysieren die duale Version der Trajektorienaustauschverfahren, d.h. wir betrachten das duale Optimierungsproblem und beschreiben die Evolution dieser Verfahren in der dualen Darstellung. Eine Hauptbeobachtung in diesem Kapitel ist, dass diese Methoden als duale Versionen der in Kapitel 2 vorgestellten Approximationsmethoden verstanden werden können. Hierfür wird zunächst gezeigt, dass das der verteilten modell-prädiktiven Regelung zugrundeliegende Optimierungsproblem direkt als konvexes, „beinahe separierbares" Optimierungsproblem in den Ausgangstrajektorien der dynamischen Systeme formuliert werden kann. Das duale Problem zu dieser neuen Problemdarstellung kann als lineares Programm mit halb-unendlichen (engl.: semi-infinite) Zwangsbedingungen formuliert werden. In dieser neuen Darstellung wird offensichtlich, dass die heuristischen Optimierungsmethoden, die auf einem Austausch von prädizierten Trajektorien basieren, auch als Cutting-Plane Optimierungsmethoden verstanden werden können. Aus dieser neuen Darstellung wird auch deutlich, weshalb die bekannten Heuristiken keine Konvergenz zur optimalen Lösung garantieren können. Speziell zeigen wir in diesem Kapitel, dass es nicht genügt, nur eine Trajektorie pro System zu speichern und auszutauschen. Basierend auf diesen Überlegungen wird gezeigt, dass der in Kapitel 2 entwickelte Algorithmus in veränderter Form auch auf das neue Problem angewendet werden kann. Der daraus resultierende Algorithmus ist wieder ein Trajektorienaustauschverfahren, hat aber eine garantierte Konvergenz zur optimalen Lösung. Wir zeigen außerdem, dass der Algorithmus Vorteile für den Einsatz in Regelungsproblemen bietet, da er schon vor der vollständigen Konvergenz zu jedem Zeitpunkt eine global zulässige (wenn auch suboptimale) Lösung liefert. Dieses Kapitel basiert auf Bürger et al. (2011a), Bürger et al. (2014), Bürger et al. (2013a) und Lorenzen, Bürger, Notarstefano, and Allgöwer (2013).

Kapitel 4: Duality and Network Theory in Cooperative Control (Dualität und Netzwerktheorie in kooperativen Regelungssystemen)

In diesem Kapitel wird eine allgemeine Dualitätstheorie für die Ausgangssynchronisation in Netzwerken passiver dynamischer Systeme entwickelt. Wir nutzen in diesem Kapitel eine verfeinerte Version der klassischen systemtheoretischen Eigenschaft Passivität, nämlich die „Gleichgewichts-unabhängige Passivität" (engl.: equilibrium-independent passivity). Ein dynamisches System hat diese Eigenschaft, wenn es für jede Ruhelage, die mit einem konstanten Eingangssignal erzeugt werden kann, passiv bezüglich dieser Ruhelage ist. Wir betrachten ein Ausgangssynchronisationsproblem für Netzwerke dynamischer Systeme dieser Art und zeigen, dass verschiedene Dualitätsrelationen existieren. Wir können diese Dualitäten identifizieren, indem wir zeigen, dass das asymptotische Verhalten der Systemvariablen optimal bezüglich einer Familie von Netzwerkoptimierungsproblemen ist. So konvergieren die Ausgänge der dynamischen Systeme zu einer Konfiguration mit minimalem „Potential" und können daher als Potentiale interpretiert werden. In einem netzwerk-theoretischen Sinn sind „Divergenzen" das duale Konzept zu Potentialen. In der Tat zeigt unsere Analyse, dass die Eingänge der einzelnen Systeme zu einer optimalen Lösung eines Netzwerkflussproblems konvergieren und somit als Divergenzen verstanden werden können. Dieses Ergebnis führt zu einer neuen Interpretation der Dualität zwischen Ein- und Ausgängen in Regelungssystemen. Wir präsentieren weitere Interpretationen der dynamischen Variablen und zeigen beispielsweise, dass die Ausgänge der Regler wie „Flüsse" in einem Transportnetzwerk verstanden werden können. Die neuen Dualitätsrelationen haben verschiedene praktische Bedeutungen. Zum einen führen sie direkt zu einer Lyapunovfunktion und somit zu einer Stabilitätsanalyse für das Regelungssystem. Zum anderen zeigen die Ergebnisse auf, wie dynamische Netzwerke entworfen werden müssen, um optimal zu arbeiten. Wir verdeutlichen die praktische Relevanz der Ergebnisse am Beispiel einer optimalen Flussregelung in einem Transportnetzwerk. Dieses Kapitel basiert auf den Veröffentlichungen Bürger et al. (2013c), Bürger et al. (2010) und Bürger et al. (2013d).

Kapitel 5: Clustering in Dynamical Networks (Clustering in Dynamischen Netzwerken)
Wir betrachten in diesem Kapitel das Phänomen, dass sich in dynamischen Netzwerken verschiedene Gruppen von Systemen ausbilden, deren Ausgänge zueinander synchron aber asynchron zu den Ausgängen von Systeme in anderen Gruppen sind. Dieses Phänomen wird weitläufig als „Clustering" bezeichnet und kann sowohl ein gewünschtes als auch ein unerwünschtes Verhalten sein. Wir zeigen zunächst, dass Clustering in passivitäts-basierten kooperativen Regelungssystemen bereits bei einer kleinen Modifikation der Regler auftreten kann. Speziell zeigen wir, dass Clustering in dynamischen Netzwerken auftritt, wenn die Kopplung zwischen den dynamischen Systemen beschränkt ist. Um das komplexe dynamische Phänomen des Clusterings zu verstehen, nutzen wir die Dualitätsrelationen und die netzwerk-theoretischen Interpretationen aus Kapitel 4. Diese geben uns verschiedene Optimierungsmethoden an die Hand, die für eine exakte Analyse des Clusterings genutzt werden können. Basierend auf diesen Optimierungsmethoden präsentieren wir kombinatorische Bedingungen dafür, dass Clustering in dynamischen Netzwerken auftritt. In einem letzten Schritt zeigen wir dann, wie die Optimierungsmethoden auch genutzt werden können, um dynamische Netzwerke strukturell zu analysieren. Wir stellen einen neuen Algorithmus vor, der, basierend auf konvexen Optimierungsmethoden, eine hierarchische Clusteranalyse der Netzwerke ermöglicht. Dieses Kapitel basiert auf den Veröffentlichungen Bürger et al. (2011c), Bürger et al. (2012c) und Bürger et al. (2013b).

Abschließend präsentieren wir eine Zusammenfassung der Arbeit sowie eine Diskussion der Ergebnisse und weiterer offener Fragen in Kapitel 6. Im Anhang werden kurz einige theoretischen Grundlagen der Konvexen Analyse und Optimierung (Anhang A), der System- und Regelungstheorie (Anhang B) sowie der Graphentheorie (Anhang C) wiederholt.

Chapter 1

Introduction

1.1 Motivation and Focus

Across almost all engineering areas there is a need for *cooperative optimization and control.* This assertion can be easily supported by various examples. In robotic networks, for example, mobile agents have to cooperate and distribute tasks in order to solve problems that no single robot can solve on its own (Behnke, 2013). In sensor networks, quantities and variables have to be estimated based on spatially distributed measurements (Stojmenovic, 2005). In smart grids, intelligent power generators and loads have to coordinate themselves such that the complete grid operates in a stable and, at best, optimal way (Zamora and Srivastava, 2010). All these examples require to select the "best" out of some available alternatives, i.e., to *optimize*, and to influence dynamical systems such that they behave as desired, i.e., to *control*. *Cooperation* is obviously required, as these engineering systems are widely expected to operate in a state that is optimal in some "global" sense. The great economic and technological potential that these novel distributed engineering systems are promising, motivates a versatile interest in cooperative optimization and control from both practitioners and theoreticians. A variety of disciplines, including computer science and control engineering, but also power systems and aerospace engineering, aim to develop conceptual and analytic frameworks for distributed, or multi-agent, systems. Within the numerous research directions related to distributed systems, two conceptual problems take an outstanding role and will be of significant importance for this thesis.

One structural task that appears repeatedly in distributed and multi-agent systems is to solve finite-dimensional optimization problems in networks of spatially distributed decision units, here also called processors. In fact, solving distributed optimization problems is important for a great share of distributed decision problems, including estimation problems in sensor networks, assignment or distribution problems in robotic networks, or dynamic control problems formulated in a model predictive control framework. All of these decision problems can be easily formulated as optimization problems. Additionally, all the named problems are inherently distributed, since the different decision makers, i.e., sensors, robots or other control systems, know only part of the problem information. Thus, as different as all these examples are, the underlying challenge is similar. The different decision makers must find an optimal solution only by exchanging data over a possibly unreliable and asynchronous communication network, such as, e.g., the internet. As this general task of solving optimization problems in a distributed manner is at the heart of many decision problems, a great share of theoretical research has been dedicated over the last years to understand and overcome the associated challenges, see, e.g., Necoara et al. (2011) for an overview. The standing challenges are, first, to identify problem classes that can be solved efficiently in a distributed way and, second, to develop explicit algorithms that solve distributed optimization problems in asynchronous networks without any coordination unit.

Another challenge in distributed systems is the analysis and design of networked dynamical systems. A situation, often encountered in modern engineering systems, is that continuously evolving dynamical systems interact with each other in some structured way, either because this interaction is inherent to the underlying problem, such as for example, in power networks, or because the control objective enforces cooperation, such as, e.g., in platooning of autonomous vehicles. Such systems should not be understood as one large monolithic system, but rather as a composition of various simpler units, i.e., they are *networks of systems.* In fact, networks of systems can show a very complex dynamic behavior, even though the single systems might have a low dynamical complexity. They require, therefore, sophisticated cooperative control strategies. The main goal of recent theoretical research is to understand how the dynamic performance of the networked systems is influenced by the interplay between the dynamic behavior of the local systems, the interaction structure, and the underlying network topology, see, e.g., the recent books Mesbahi and Egerstedt (2010), and Bai et al. (2011). In order to be able to reach this ultimate goal, one needs to understand first under which condition networks of dynamical systems behave nicely and when their behavior is optimal with respect to some desired performance criterion. Consequently, computational analysis tools have to be developed that allow an effective and exact analysis of the network behavior. Only such analysis methods will open the way for an optimal control design for networked systems as well as an understanding of other more complex dynamic behaviors.

This thesis addresses different problems in cooperative optimization and control, related to the challenges described above. A first methodological idea, that appears repeatedly throughout this thesis, is the one of *approximation.* Since it is often difficult to handle decision problems directly in a distributed setting, we employ here an approximation idea and solve repeatedly a sequence of simpler approximations of the problem. By refining the approximations in a suitable way, this proceeding will eventually provide the desired solution of the original problem. The approximation idea is heavily exploited in this thesis and provides the basis for a family of distributed algorithms to solve a wide range of problems, from robust optimization to distributed predictive control problems.

An even more important concept in this thesis is *duality.* Duality is the mathematical manifestation of the idea that there are always two sides to a problem.[1] It is often very insightful to study a problem from both, the original (so-called primal) as well as from the dual perspective, and this will be a recurrent theme of this thesis, leading the way to various important conclusions. For example, duality will allow us to establish a theoretical and methodological connection between problems such as position estimation in sensor networks and cooperative predictive control. As we will see, the dual problem of a cooperative optimal control problem is structurally similar to many other important distributed optimization

[1]For example, finding the minimum distance of a point to a line is dual to the problem of finding the maximal distance to planes through that line. In fact, if one of the two problems is solved, the other problem is also implicitly solved.

problems. Already this simple observation will lead the way to an efficient algorithm design.

Adding to this, duality turns out to be of major importance for networked systems and in particular for cooperative control problems. A canonical model in cooperative control is the output agreement problem, where different dynamical systems have to be controlled such that their outputs behave identically. Output agreement problems appear in a variety of distributed control problems, including formation control, routing control etc., and have been in the focus of control theoretic research over the last decade. As a main theoretical contribution of this thesis, we establish several duality relations between the dynamic variables of a passivity-based cooperative control framework. We present a general model for output agreement problems involving passive dynamical systems and show that the asymptotic behavior of the network is inherently related to several convex optimization problems. In fact, we show that the inputs and outputs of the dynamical systems are dual variables to each other, in the standard sense well-known from convex optimization.

While these duality relations are interesting in their own right, they also turn out to be of significant practical relevance. They open the way to understand more complex variations of the problem. For example, the duality relations help to understand the more challenging dynamic phenomenon of "clustering" (or partial output agreement). Clustering refers to the situation, where only groups of dynamical systems reach an agreement on their outputs. It can therefore be understood as a generalization of output agreement and from a control perspective it is desirable to understand the mechanisms leading to clustering. However, clustering is a complex dynamic behavior that is non-trivial to be understood from a purely primal perspective. We show in this thesis, that a combination of the primal and dual perspective is the key for understanding clustering. In fact, the proposed duality relations will lead the way to constructive analysis tools.

1.2 Contributions and Organization

This thesis contributes towards a general theory of cooperative optimization and control. It addresses the general questions *"how can decisions be taken efficiently in a distributed environment?"* and *"how can cooperative dynamical systems be designed to work optimally together?"* We describe in the following the main contributions of the thesis and the organization as well as the detailed contributions of each chapter.

Main Contributions

First, a broad framework for cooperative convex and robust optimization is established. We identify a general class of distributed optimization problems that can be efficiently handled in peer-to-peer processor networks with asynchronous communication. The considered problem class relates to an abstract problem formulation, where processors have to agree on the optimizer of a linear cost function over the intersection of local convex constraint sets. Contributing towards a general theory, we propose here an abstract distributed algorithm for solving this general problem class, without being restricted to a certain representation. In particular, we propose and analyze an algorithm that uses sequential polyhedral approximations of the constraint sets. The algorithm is such that polyhedral approximations are exchanged between the processors, and the approximations are improved by a local generation of so called "cutting-planes". We show that the algorithm performs correctly and has an inherent robustness against communication or processor failures. The abstract algorithm is shown to be applicable to several important problem representations. We show how the general algorithm can solve semi-definite or robust optimization problems. A variety of relevant decision problems are shown to be contained in the considered general problem class.

Second, optimal control problems in a model predictive control framework are considered. These control problems can be associated with a finite-dimensional distributed optimization problem. An established way to handle those problems distributedly is to use trajectory exchange methods, where processors exchange their predicted trajectories between each other. We provide here an interpretation of those trajectory exchange methods as dual versions of the distributed polyhedral approximation method. In particular, we show that from an optimization perspective a trajectory exchange method is equivalent to a polyhedral approximation exchange method for the dual problem. This dual interpretation explains why the basic trajectory exchange methods, as known from the literature, cannot solve cooperative control problems exactly. Following this, the polyhedral approximation interpretation serves as the basis for the design of an improved trajectory exchange algorithm, that performs provably correct in asynchronous communication networks and will compute the central optimal solution, if it is performed sufficiently long.

These first two main contributions of the thesis are related to the problem of solving finite dimensional optimization problems within networks of distributed processors. The remaining main contributions aim towards analyzing the behavior of interacting dynamical systems and towards developing optimal network design methods.

As a third main contribution of this thesis, a duality theory for passivity-based cooperative control is presented. A cooperative control framework is considered, where distributed dynamical systems are controlled to reach output agreement. A variant of the classical system property passivity, i.e., equilibrium independent passivity, is considered to establish a general problem description that generalizes various existing setups. It is shown that the dynamical network admits asymptotically a certain optimal behavior. To show this, we establish an explicit connection between the dynamical variables and a family of network optimization problems. This "inverse optimality" result unveils several duality relations between the variables. In particular, we show that the outputs minimize an "optimal potential problem" and admit therefore an interpretation as "potentials", in the network theoretic sense defined in (Rockafellar, 1998). We show, furthermore, that the inputs of the control system can be understood as the dual variables to these potential variables, that are named "divergence" in network theory. In this light, the results provide a theoretical understanding of the mechanisms underlying the output agreement problem. Additionally, they lead the way to constructive methods for a dynamical network analysis and design, as they allow us to use classical optimization methods to analyze the cooperative dynamical behavior.

Finally, as a fourth contribution, the phenomenon of "clustering" in dynamical networks is investigated and explained in the proposed duality framework. We consider a variation of the output agreement problem, where the interaction structure between the dynamical systems is slightly modified. We show that under the given modification, the network does no longer reach complete agreement, but rather only groups of systems agree. We explain this complex dynamic phenomenon using the previously introduced duality interpretations. By studying the dynamical network from a combined primal and dual perspective, we will provide an explanation of the clustering phenomenon. Additionally, since the duality relations allow us to connect the asymptotic network behavior to a family of network optimization problems, we also directly obtain tools, i.e., standard optimization methods, for analyzing and predicting the clustering behavior.

Summarizing, the four predominant contributions of this thesis are the following:

(i) We propose an algorithm for distributed convex and robust optimization in peer-to-peer processor networks with asynchronous communication.

(ii) We present a duality interpretation of trajectory exchange methods in predictive control as polyhedral approximation methods; based on this interpretation we present an improved trajectory exchange optimization algorithm.

(iii) We present a complete duality theory for passivity-based cooperative control.

(iv) We derive optimization-based methods for a clustering analysis of dynamical networks.

Organization

The organization of the thesis reflects the four major contributions. The main part of the thesis is organized in four chapters and in each chapter one of the main contributions is presented. In more detail, the contributions of the single chapters are as follows.

Chapter 2: Polyhedral Approximation Methods for Cooperative Optimization In this chapter, a general framework for distributed convex and robust optimization in peer-to-peer processor networks with asynchronous communication is developed. First, an overview of the considered network model and several complexity notions are presented. A general formulation of distributed convex optimization problems is introduced, where processors are assigned general convex constraint sets. The processors have to cooperatively agree on an optimizer of a linear objective function over the intersection of the constraint sets. We propose in this chapter a distributed optimization algorithm to solve the proposed class of optimization problems. To prepare the algorithm definition, we first discuss some preliminary results on unique solutions to linear programming problems. We discuss, how linear programs can be efficiently solved according to a unique solution rule, i.e., we discuss the lexicographic minimal solution and the minimal 2-norm solution of linear programs. Based on this preliminary result, we formally introduce a distributed optimization algorithm. The algorithm we propose relies on a polyhedral approximation or cutting-plane idea and is consequently named *cutting-plane consensus algorithm*. We present the algorithm first in a general formulation that is independent of the specific representation of the constraint sets. Then, for the general algorithm description, we prove correctness as well as robustness against communication or computation faults. Following this general discussion, we specify the algorithm for certain problem representations. The first specific problem representation we consider in this chapter relates to convex optimization problems with inequality or semi-definite constraints. We show that the algorithm can be directly applied to solve problems of this form, by using subgradients to define the approximation. In this context, it turns out that the algorithm becomes a distributed simplex algorithm when the original optimization problem is already linear. To illustrate the practical relevance of the method, we discuss, how the algorithm can solve a localization problem in sensor networks.

The second problem representation, for which we design a new version of the general algorithm, is robust optimization with uncertain constraints. We consider optimization problems with uncertain constraints, where cooperatively solutions that are feasible for any possible representation of the uncertainty have to be computed. With only a small modification, i.e., by using a certain cutting-plane oracle, the proposed algorithm is able to solve robust optimization problems in peer-to-peer processor networks. We classify several representations of the uncertain constraints that can be solved efficiently by our algorithm. Finally, we present a computational study for robust linear programming that compares the performance of the algorithm to other distributed optimization methods.

The results presented in this chapter are partially based on Bürger et al. (2011b), Bürger et al. (2012b), Bürger et al. (2012a), Bürger et al. (2014). The main contributions of this chapter can be summarized as follows:

- We propose a distributed algorithm for cooperative optimization in asynchronous peer-to-peer networks.
- We prove the correctness and robustness of the novel algorithm.
- We specify the algorithm to important problem classes, such as semi-definite or robust optimization problems.

Chapter 3: Dual Cutting-Plane and Trajectory Exchange Optimization

In this chapter, we consider *trajectory exchange methods*, which are studied intensively in distributed model predictive control, and show that they can be interpreted as (inconsistent) dual versions of the cutting-plane consensus algorithm. We consider therefore a finite horizon optimal control problem involving several linear discrete-time dynamical systems, with their outputs being coupled by some constraints. The cooperative control problem is now the optimal coordination of the group of dynamical systems. That is, the control inputs have to be chosen such that the resulting trajectories satisfy the coupling constraints and optimize the cost of the complete group. We show first that the predictive control problem can be equivalently represented as an almost separable convex optimization problem in the output trajectories. The problem representation in the output trajectories is then the basis for the discussions in this chapter. A key observation we make is that the partial Lagrange dual of the cooperative predictive control problem can be expressed as a linear optimization problem with semi-infinite constraints.

Next, we consider a trajectory exchange method proposed in the distributed model predictive control literature and analyze it in the dual formulation. Considering the dual representation, it turns out that the trajectory exchange method can be seen as a (non-convergent) variant of a distributed cutting-plane optimization method. In particular, we show that in trajectory exchange methods processors store and exchange polyhedral approximations of the *dual* feasible set. This dual interpretation explains directly why the trajectory exchange method does not converge to a central optimal solution. We show that for convergence to the centralized optimal solution it is necessary to store and exchange more than one trajectory per system.

To overcome this issue and to design a convergent algorithm, we show how the previously proposed cutting-plane consensus algorithm can be redesigned to solve this class of cooperative control problems. In a first step, we apply the algorithm to the dual problem representation of a general almost separable convex optimization problem. We discuss the convergence of the algorithm and show that at any time instant a primal feasible solution can be reconstructed, even before the algorithm has converged. As the algorithm converges, the reconstructed primal solution becomes optimal. The algorithm applied to solve the dual has various interpretations when it is analyzed from the primal perspective. In the first place, we show that the algorithm has an interpretation as a fully distributed variant of the classical *Dantzig-Wolfe Decomposition*, and is an inner linearization method for possibly non-linear programs. Complementing this observation, we see that, in the context of cooperative optimal control, the algorithm is in fact a trajectory exchange method, where the single systems store and exchange a set of feasible trajectories and update this set by planning locally new output trajectories. The algorithm is therefore in the spirit of established trajectory exchange methods. In fact, the algorithm keeps the advantageous properties of trajectory exchange methods, as it provides, for example, a feasible solution of the problem already with very little communication between the systems. Additionally, the algorithm is cooperative in the sense that if more communication is allowed, it will also recapture the central optimal solution.

The results of this chapter are based on Bürger et al. (2011a), Bürger et al. (2014), Bürger et al. (2013a), Lorenzen, Bürger, Notarstefano, and Allgöwer (2013). The main contributions of this chapter are:

- We present the dual formulation of the cooperative optimal control problem as semi-infinite linear program.
- We provide an interpretation of trajectory exchange methods as dual cutting-plane algorithms.
- We propose an improved trajectory exchange optimization method, based on the cutting-plane consensus algorithm.

Chapter 4: Duality and Network Theory in Cooperative Control
In this chapter, we present a general *duality theory for output agreement* problems. We consider a classical continuous time cooperative control framework, involving a network of dynamical systems that satisfy a certain passivity property. The cooperative control problem is to reach an agreement on the output variables, using only nearest neighbor interactions. In the set-up considered here, the systems are assumed to be *equilibrium independent passive* systems. Equilibrium independent passivity is a variation of the classical passivity concept. Roughly speaking, a dynamical system is equilibrium independent passive, if it is passive independent of the equilibrium point to which it is regulated. A class of non-linear couplings between the systems, using only relative output information between neighboring systems, is proposed that achieves output agreement asymptotically. The proposed cooperative control framework is a generalization of several well-established cooperative control problems.

We show that the output agreement problem in such dynamical networks admits several duality relations. To identify the duality relations, we connect the asymptotic behavior of the dynamical network to two dual-pairs of optimization problems. In a first step, we consider only the plant level and the structural constraint on the control input resulting from the network structure, without specifying the control law yet. The possible output agreement state turns out to be the minimizer of a certain optimal potential problem. This provides an interpretation of the output variables as potentials, in the network theoretic sense of Rockafellar (1998). Dual to this, we show that the inputs have an interpretation as divergences and that their steady state value is determined by an optimal flow problem. Building upon this observation, a certain control structure for the output agreement problem is chosen and the asymptotic behavior of the controller is connected again to a dual pair of network optimization problems, i.e., an optimal flow and an optimal potential problem. We provide an interpretation of the controller output as optimal flows in a network. Dual to this, we give an interpretation of the controller states as tensions. As a first practical implication, the presented duality relations provide directly a Lyapunov function and a stability proof for the dynamical network. We illustrate the practical usefulness of the duality relations on an exemplary control problem and show that the duality relations lead directly the way to a distributed controller design for optimal distribution control in dynamic inventory systems.

The results of this chapter are partially based on Bürger et al. (2013c), Bürger et al. (2010), Bürger et al. (2013d). The contributions can be summarized as follows:

- We introduce the concept of equilibrium independent passivity into a general framework for cooperative control.
- We identify several duality relations in passivity-based cooperative control.
- We establish an explicit connection between passivity-based cooperative control and network optimization theory.

Chapter 5: Clustering in Dynamical Networks

In this chapter, we study clustering in dynamical networks. We call clustering the phenomenon that groups of nodes in a network agree with their outputs, while being in disagreement with other nodes. Clustering is a generalization of output agreement and we study the problem here from a cooperative control perspective. In this chapter, we build upon the passivity-based cooperative control framework studied in Chapter 4. We show, in particular, that a simple modification of the controller structure leads to a clustering of the network. In this sense, we derive a model for dynamic clustering directly as a variation of the classical cooperative control framework. In order to analyze and understand the clustering phenomenon, we exploit the duality relations established previously.

The discussion begins by considering a static network optimization framework. We consider a dual pair of network optimization problems, one optimal flow problem and one optimal potential problem. We show that if in the optimal flow problem some constraints on the flows are active, then the optimal potential (dual) variables have a clustered structure. The clustering structure of the potential variables cannot be directly seen in the original formulation of the optimization problems. However, it is revealed by an optimization problem that is "between" the two original problems. In particular, we consider a saddle-point problem involving the potential and the flow variables. The saddle-point problem is then used to analyze the clustering structures in the static, network theoretic sense. Following this static discussion, we turn our attention to the dynamical networks. We use the explicit connection between the network optimization framework and the passivity-based cooperative control framework to prove that a certain class of dynamical networks with bounded interaction rules exhibits clustering in its asymptotic behavior. In fact, by exploiting the established duality relations, we propose a dynamic counterpart of the flow constraints and the clustered potential variables. As the outputs of the dynamical systems in the cooperative control framework are intimately related to the potential variables, a clustering dynamical network model can be defined by integrating the flow constraints into the cooperative control framework. It turns out that the flow constraints correspond to controllers with saturated output functions. In this way, we can characterize a class of dynamical networks that exhibit asymptotically a clustering behavior. By exploiting again the duality relations between the system variables, we can derive a Lyapunov analysis to prove the convergence.

Finally, we use the cluster synchronization problem to define an algorithm for detecting hierarchical clustering structures in dynamical networks. As the static network optimization problems can be solved efficiently, they provide an efficient tool to identify clustering structures in networks. In a first step, we investigate the combinatorial properties of the appearing clustering structures. We show that the clustering structures satisfy specific optimality criteria, based on an interpretation of the network as a weighted graph. Following this analysis, we propose a computationally attractive algorithm, based on convex optimization, to identify hierarchical clustering structures in networks. We show that our method generalizes existing network partitioning methods and can, for example, be used for structure identification in power networks.

The results of this chapter are based on Bürger et al. (2011c), Bürger et al. (2012c), Bürger et al. (2013b). The main contributions can be summarized as follows:

- A saddle-point analysis method for the identification of clustering structures in constrained optimal flow problems is proposed.
- We identify and analyze a class of cooperative dynamical networks that exhibit asymptotically clustering.
- We propose a hierarchical clustering algorithm based on convex optimization.

Complementing these four main chapters of the thesis, a concluding discussion and an outlook on open problems are presented in Chapter 6. For the sake of completeness, the theoretical foundations are presented in the appendix. Some fundamentals on convex analysis and convex optimization are reviewed in Appendix A. The basics of dynamical systems and control theory are presented in Appendix B, before some basic graph theoretic concepts are provided in Appendix C.

Chapter 2

Polyhedral Approximation Methods for Cooperative Optimization

2.1 Introduction

It is well known, that many important decision problems in engineering systems can be formulated as optimization problems. Since many of those decision problems have also a distributed nature, there is an increasing interest in algorithms that can solve optimization problems in a distributed way. The basic problem is to find an optimal decision variable over some constraint set within a group of distributed processors, each having knowledge of a part of the problem, by exchanging data over a communication network. If all the processors in such a system are identical (i.e., "peers") and none of the processors takes a coordinating role, we say that the algorithm performs in a *peer-to-peer network*. Clearly, algorithms that can perform in peer-to-peer networks must satisfy certain requirements that arise from the communication structure. Some of the requirements on the distributed algorithms in peer-to-peer networks are the following, see also Hendrickx et al. (2011):

Identical Processors: All processors in the network are exactly identical and any two processors with the same input data will perform exactly the same computations and obtain the same output. This assumption does not imply that all processors have the same (synchronized) clocks or perform their computations with the same speed.

Bounded Memory: The processors are small computational units, embedded in possibly very large network structures. Each processor has only a bounded memory available and, in order to achieve scalability, the amount of data assigned by the algorithm to a processor must be independent of the overall network size.

Lack of Global Knowledge: There is no global knowledge. That is, no variable in the system is known to all processors and no single processor has knowledge about the complete problem. In this way, no processor can solve the problem independently but the solution can only be obtained through communication with other systems.

The origins of distributed, or peer-to-peer, optimization can be found in the *parallel optimization* (Bertsekas and Tsitsiklis, 1997) or *large-scale optimization* (Lasdon, 2002) literature. The objective of parallel optimization is to speed up optimization algorithms by performing different computational steps in parallel on multi-processor systems. Most of the existing parallel algorithms are designed for processor architectures, where the different units can all access one common memory. In this way, the processors can exchange large amounts of data very efficiently and fast. While parallel algorithms have proven very successful from a computational perspective, it is obvious that they are not designed to handle distributed optimization in peer-to-peer networks, where the processors are spatially distributed and the communication is eventually unreliable. Large scale optimization, on the other hand, focuses on the decomposition of possibly very large optimization problems into several smaller problems. Those decomposition algorithms often represent a large but structured optimization problem as several sub-problems and one coordinating master program. Usually, the master program is the critical component for solving a large-scale optimization problem. This master program centrally collects data from all sub-problems and provides all systems simultaneously with updated variables. Well-known representations of such large-scale optimization, or decomposition methods are the Dantzig-Wolfe decomposition (Dantzig and Wolfe, 1961) or dual gradient algorithms (Bertsekas and Tsitsiklis, 1997). Considering application areas for distributed optimization, as described above, some drawbacks of the parallel or large scale optimization algorithms become obvious. First, the existence of a central master problem contradicts the concept of peer-to-peer optimization, since the one processor solving the master program must be inherently different from the processors solving the sub-problems. A failure of the central (i.e., master) processor will lead to a complete failure of the optimization algorithm. The second problem refers to the timing issue. In fact, a master/sub-problem structure requires a *synchronization* between all processors in the network. More precisely, all processors solving the sub-problems have to finish their computations at the same point in time, or have to waste time by waiting for other processors. The master program also needs to be synchronized with the complete network, as it has to know when all other processors finished their local computations. One can easily see the drawbacks of such approaches, when the problems have to be solved by a very large number of processors that are spatially distributed across a wide area and have to use an unreliable communication network. In such a set-up, algorithms without any coordination unit are highly desirable. This observation has led to a significant academic interest over the last decade and consequently to a rich variety of novel peer-to-peer optimization algorithms. We summarize next some of the most important research directions related to distributed optimization in peer-to-peer networks. As the literature on distributed optimization is vast and steadily growing, this review has no claim to completeness.

Initially major attention was given to asynchronous *distributed subgradient methods*, see e.g., Tsitsiklis et al. (1986), Nedic and Ozdaglar (2009). The basic problem formulation in distributed subgradient optimization, see e.g., Nedic and Ozdaglar (2009), is that each processor is assigned an individual convex objective function, and all processors have to agree on one decision variable that minimizes the sum of all objective functions. Such distributed subgradient algorithms connect very elegantly to the well-known agreement or consensus problem (Jadbabaie et al., 2003), (Moreau, 2005). For optimization problems that have inherently a networked structure, distributed subgradient methods can be applied to the dual problem. Due to this fact, distributed primal and dual subgradient algorithms became important tools in network utility maximization and have been intensively studied in the communication networks literature, see Low and Lapsley (1999), Low et al. (2002). Combined with projection operations, subgradient methods can also solve constrained optimization problems (Nedic et al., 2010), (Low and Lapsley, 1999). Distributed subgradient algorithms are today fairly well understood (Xiao and Boyd, 2006), (Johansson et al., 2009) and are already used in distributed control problems, e.g., in distributed model predictive control, see Doan et al. (2011), Giselsson and Rantzer (2010). The main advantage of subgradient algorithms is that they meet the requirements of distributed peer-to-peer systems. In particular, the algorithms perform in asynchronous communication networks and require only a nearest neighbor communication. However, the algorithm need in general a balanced communication graph.

Building upon these results for distributed optimization with subgradient methods, the research scope has been widened in the last years. Several novel distributed algorithms were explored with the intention to improve the convergence speed or to address novel classes of optimization problems. In this context, *distributed Newton methods* have found quite some attention. A distributed newton method for network utility maximization with strictly convex objective functions has been proposed in Zargham et al. (2011b) and Zargham et al. (2011a). The algorithm exploits the inherently distributed structure of the underlying optimization problem and can ensure locally a quadratic convergence rate. For unconstrained quadratic or strongly convex optimization problems a Newton-Raphson consensus is proposed in Zanella et al. (2011). We want to point out that both algorithmic schemes are distributed in the sense that no central coordination unit is required, while only the latter is asynchronous.

Following the distributed newton methods, recently also distributed variants of *Alternating Direction Methods of Multipliers* (ADMM) have been proposed. ADMM is basically a primal/dual descent method for an augmented Lagrangian function. In particular, the standard Lagrangian function is augmented with a quadratic penalty function for linear constraints. This quadratic extension makes ADMM flexible and applicable to most convex optimization problems, i.e., the method does not require strict convexity of the objective functions. Much attention was given to distributed variants of Alternating Direction Method of Multipliers (ADMM). The recent literature on this algorithm is too vast to be reviewed here, and we discuss only some of the recent developments. In Boyd et al. (2010), ADMM was investigated for machine learning problems. A distributed implementation is achieved by replacing a central update step of the algorithm with a consensus algorithm. In the earlier work of Schizas et al. (2008), a distributed ADMM scheme was proposed in the context of distributed estimation. The method does not require a central coordination step or a consensus algorithm, but requires a static and undirected communication graph. Recently, a convergence rate analysis was presented in Wei and Ozdaglar (2012), where the algorithm performs distributed using a Gauss-Seidel iteration scheme.

Complementing the gradient-based approaches, an alternative algorithmic approach has been proposed in Notarstefano and Bullo (2007) and Notarstefano and Bullo (2011). There, a fully distributed, asynchronous algorithm for *abstract programs* has been proposed. Abstract optimization problems, see e.g. Gärtner and Welzl (1996a) or Amenta (1993), are optimization problems with linear objective functions, where the optimal solution is fully defined by a small number of constraints. In Notarstefano and Bullo (2011), a simple distributed algorithm is proposed to solve abstract programs, if the information about the constraints is distributed to different processors. The main advantage of the algorithm proposed in Notarstefano and Bullo (2011) is its low communication requirement. In fact, the algorithm does not require a synchronous communication. It does also not require a balanced communication graph, but only a jointly connected graph. The idea of Notarstefano and Bullo (2007) has been further refined for convex programs with a large number of similar constraints in Carlone et al. (2012). The distributed optimization algorithms presented in this thesis are inspired by the results of Notarstefano and Bullo (2011), but address the problem from another direction, in the sense that they fully exploit the problem structure and do not require a common representation of the constraints.

2.2 Distributed Algorithms in Peer-to-Peer Networks

We introduce now the conceptual framework for algorithms performing in peer-to-peer networks. First, we describe the general model of the communication network, then we present a general description of distributed algorithms. Finally we present some complexity notions that are used to quantify the algorithms performance.

2.2.1 Communication Network Model

To start the discussion, we formalize a model for the distributed computation system. We consider in particular a standard model for distributed processor systems, as introduced, for example, in Bullo et al. (2009).

Definition 2.1 (Communication Network). *A communication network is a (possibly time-dependent) digraph $\mathcal{G}_c(t) = (\mathbf{V}, \mathbf{E}_c(t))$, where*

- $t \in \mathbb{Z}_{\geq 0}$ *is the universal (slotted) time; and*
- $\mathbf{V} = \{1, \ldots, n\}$ *is a set of unique identifiers; and*
- $\mathbf{E}_c(t)$ *is the (time-dependent) set of directed edges over the vertices* $\{1, ..., n\}$*, called the communication links.*

In general, it is not necessary that all processors know the value of the universal time t. This universal time determines which communication structure, i.e., which graph $\mathcal{G}_c(t)$, is currently active. The time-dependent graph $\mathcal{G}_c(t)$ is called *communication graph.* It has the fixed node set $\mathbf{V}$ and the time-varying edge set $\mathbf{E}_c(t)$. It models the communication in the sense that at time t there is an edge from node i to node j in $\mathbf{E}_c(t)$ if and only if processor i transmits information to processor j at time t. The time-dependency of the communication graph is a simple way to model an unreliable communication, that is, e.g., due to package losses or wireless communication with time-dependent connections between the different units.

For such time-dependent networks, we need to refine some of the graph theoretic concepts.[1] The time-varying set of outgoing (incoming) *neighbors* of node i at time t, denoted by $\mathcal{N}_O(i,t)$ ($\mathcal{N}_I(i,t)$), are the set of nodes to (from) which there are edges from (to) i at time t. Also the notion of connectivity needs to be refined for time-varying graphs. A static digraph is said to be *strongly connected* if for every pair of nodes (i, j) there exists a path of directed edges that goes from i to j. For the time-varying communication graph, we rely on the notion of a uniformly jointly strongly connected graph.

Definition 2.2 (Uniform Joint Connectivity). *A communication network $\mathcal{G}_c(t)$ is said to be uniformly jointly connected, if there exists a positive and bounded duration T_c, such that for every time instant $t \in \mathbb{Z}_{\geq 0}$, the digraph $\mathcal{G}_c^{t+T_c}(t) := \cup_{\tau=t}^{t+T_c} \mathcal{G}_c(\tau)$ is strongly connected.*

[1]For a general introduction to graph theory see Appendix C.

Loosely speaking, this condition requires that every processor communicates at least once in any time interval of length T_c, and the information it sends out can reach any other processor within a finite time. We want to point out, that we require the existence of a bounded duration T_c, but we do not require to know this time. In principle, T_c might be arbitrarily large, as long as it is finite. This is a fairly weak notion of connectivity, as it does not require the communication graphs to be connected at any single time instant.

2.2.2 Distributed Algorithms

The processors in the network run *distributed algorithms*. We provide next a formal introduction to distributed algorithms. In what follows, the notational convention is employed that the superscript $[i]$ denotes that a quantity belongs to processor $i \in \mathbf{V}$.

Definition 2.3 (Distributed Algorithm). *A distributed algorithm on a communication network consists of the sets*

1. Σ *called the* communication alphabet, *with its elements being called messages;*
2. $\mathbb{W}$ *called the set of* processor states $w^{[i]}$*;*

and the two maps

1. $\texttt{MSG} : \mathbb{W} \times (1, \ldots, n) \to \Sigma$, *called* message function*; and*
2. $\texttt{STF} : \mathbb{W} \times \Sigma^N \to \mathbb{W}$, *called the* state transition function.

The definition we present here restricts the discussion to algorithms, where all processors are exactly identical and perform the same computations. That is, the set of processor states as well as the maps MSG and STF are identical for all processors. In Bullo et al. (2009), these algorithms are called *uniform* distributed algorithms. As we restrict the discussion here to those algorithms, we will omit the adjective uniform in the following.

A distributed algorithm evolves according to the following iterative procedure (see Figure 2.1). The state of each processor at time t is the variable $w^{[i]}(t) \in \mathbb{W}$. The algorithm starts at $t = 0$ and each agent initializes its state to some $w^{[i]}(0) \in \mathbb{W}$. Then, every agent i performs two actions repeatedly:

1. it computes, based on its current state, a message using the map $\texttt{MSG}(w^{[i]}(t))$, and transmits this message to all its out-neighbors $j \in \mathcal{N}_O(i, t)$ in the communication graph $\mathcal{G}_c$;
2. whenever it receives information from its in-neighbors $\mathcal{N}_I(i, t)$, it updates its state $w^{[i]}(t)$ according to the state transition function, i.e.,

$$w^{[i]}(t+1) = \texttt{STF}\Big(w^{[i]}(t), \bigcup_{j \in \mathcal{N}_I(i,t)} \texttt{MSG}(w^{[j]}(t))\Big).$$

Such an algorithm is said to run on a synchronous communication network, as all processors perform their communication and computations in parallel. However, as we allow the communication graph $\mathcal{G}_c(t)$ to be time-varying and only jointly connected, processors can remain idle during such communication rounds. In this way, this algorithmic structure models, to some extent, also asynchronous distributed algorithms.

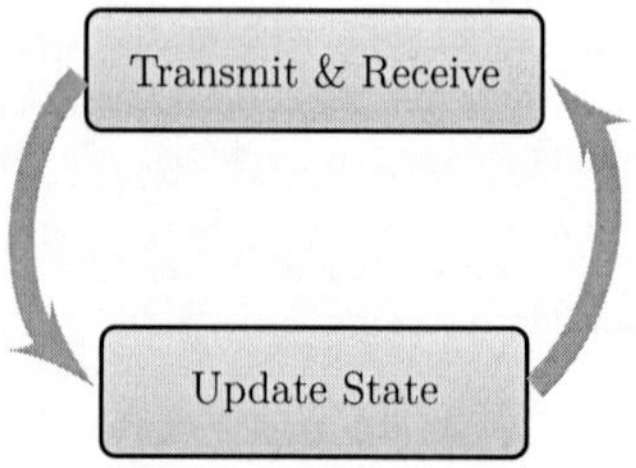

Figure 2.1: Iterative structure of a distributed algorithm.

We say in the following that an algorithms is *correct*, if the states converge to a solution of the considered problem, which is going to be here an optimization problem. Convergence can either happen asymptotically, such that the agents reach the optimal solution only if the algorithm is performed for an infinitely long time, or in finite time, such that all agents know the correct solution after a finite number of iterations. In any case, we do not necessarily require that the agents detect convergence in a distributed way.

2.2.3 Complexity Notions

We introduce now some complexity notions to quantify the performance of the distributed algorithms. As we do not necessarily require that the processors stop running the algorithm in a finite-time, we say in the following that the algorithm terminates when a centrally defined criterion is met by all processors in the network. The following notions of complexity are used for evaluating the performance of an algorithm, see e.g. Bullo et al. (2009). A first performance measure relates to the maximal number of iterations required by the algorithm until termination.

Definition 2.4 (Time Complexity). *The (*worst-case*) *time complexity *of a distributed algorithm is the maximum number of rounds required by the distributed algorithm among all allowable initial states until termination.*

Note that this notion of time complexity assumes that all local computations are preformed in a unit time. We will later on consider also the *average time complexity* as the average number of rounds required by the algorithm over all allowed initial conditions. The second relevant measure considers the maximal amount of data a processor performing the algorithm has to store.

Definition 2.5 (Space Complexity). *The (*worst-case*) *space complexity *of a distributed algorithm is the maximum amount of data stored by a processor executing the distributed algorithm among all processors and among all allowable initial states until termination.*

The space complexity is usually measured in bits. For a scalable algorithm, it is desirable that the space complexity of an algorithm does not grow with the number of processors in the network, but only slowly with the maximum degree of a processor in the communication graph. Finally, we want to measure how much usage the processor makes of the communication network. This is measured with the communication complexity.

Definition 2.6 (Communication Complexity). *The (*worst-case*)* communication complexity *of a distributed algorithm is the maximum amount of data transmitted over the entire network during the execution of the algorithm among all allowable initial states until termination.*

To quantify the complexity of the algorithm, we will use repeatedly the *Bachman-Landau symbols* (see e.g., Bullo et al. (2009)): For $f, g : \mathbb{N} \mapsto \mathbb{R}_{>0}$, we say that $f \in \mathcal{O}(g)$ (resp., $f \in \Omega(g)$) if there exists $n_0 \in \mathbb{N}$ and $K \in \mathbb{R}_{>0}$ (resp., $k \in \mathbb{R}_{>0}$) such that $f(n) \leq Kg(n)$ for all $n \geq n_0$ (resp., $f(n) \geq kg(n)$ for all $n \geq n_0$).

2.3 The Cutting-Plane Consensus Algorithm

We propose now a novel distributed optimization algorithm to solve a general class of convex optimization problems. Motivated by several important applications, we consider a broad distributed optimization framework, where each processor has knowledge of a convex constraint set, and a linear cost function has to be optimized over the intersection of these sets. The algorithm we propose uses the idea of polyhedral outer approximations of the constraint set. Processors running the algorithm generate and exchange repeatedly linear constraints, also called *cutting-planes*, that approximate the constraint sets. Instead of solving the convex optimization problem directly, the processors solve a sequence of linear programs and converge eventually to the optimal solution of the original problem. Our algorithm is consequently named Cutting-Plane Consensus. As a main contribution, we refine the idea of polyhedral approximation here in such a way that it can be performed in distributed processor networks.

In particular, the algorithm we propose requires the processors to store and exchange only a *small and fixed* number of linear constraints. The memory requirements of the algorithm are therefore fixed and are not growing during the evolution of the algorithm. We show that the amount of data stored by a processor is in many applications independent of the network size. The bounded memory requirements can only be achieved by using a special solution structure for the linear programs that approximate the original convex program. It turns out that a critical requirement for the algorithm to work is that the linear programs are solved according to a *unique solution* rule. In general, linear programs can have a continuum of optimal solutions. However, it will turn out that for our algorithm to work correctly, it is important that there is a unique criterion for selecting an optimal solution, if it is not unique. We discuss here two such selection rules, i.e., the lexicographic minimal solution and the minimal 2-norm solution. Both selection rules have different advantages, with the minimal 2-norm solution being computationally slightly more attractive. An additional advantage of the novel algorithm is that it imposes very little requirements on the communication network. In fact, we will show that the sole requirement the algorithm imposes on the network is *joint strong connectivity.* This is possibly the weakest assumption on a communication network one can expect in the context of distributed optimization.

We present in this section the theory of the cutting-plane consensus algorithm. The chapter is organized as follows. First, we discuss some preliminary results concerning unique solutions for linear programs. We discuss in particular the lexicographic minimal solution and the minimal 2-norm solution and provide computational tools to find such solutions. Then, we present the cutting-plane consensus algorithm in a general form and discuss its properties. Following this we present a detailed technical analysis of the algorithm's properties and prove its correctness. We also show how exceptional situations, such as infeasibility or unboundedness, can be handled. Based on the general theory developed in the present section, we will formulate later on more specific representations of the algorithm for different problem formulations, such as, e.g., inequality constrained problems (Section 2.4) or robust optimization problems (Section 2.5). As a preview, we also want to mention that the algorithm is applicable to almost separable optimization problems, where it becomes a fully distributed, asynchronous version of the Dantzig Wolfe decomposition and has an interpretation as a trajectory exchange method (Chapter 3).

2.3.1 General Problem Formulation

We consider here a general class of distributed optimization problems. We consider a network of processors, as defined in Section 2.2, consisting of a set of processors $\mathbf{V} = \{1, \dots, n\}$ exchanging data over a communication network. The basic assumption we make is that each processor $i \in \mathbf{V}$ has knowledge of a convex and closed constraint set $\mathcal{Z}_i \subset \mathbb{R}^d$. The decision problem, the processors have to solve cooperatively, is to agree on a decision vector $z \in \mathbb{R}^d$ maximizing a linear objective over the intersection of all sets $\mathcal{Z}_i$. That is, the processors have to solve the following class of distributed convex optimization problems

$$\begin{aligned} \text{maximize} \quad & c^\top z \\ \text{subj. to} \quad & z \in \bigcap_{i=1}^{n} \mathcal{Z}_i. \end{aligned} \tag{2.1}$$

We denote the overall feasible set in the following as $\mathcal{Z} := \bigcap_{i=1}^{n} \mathcal{Z}_i$. For clarity of presentation, we assume, if not explicitly stated differently, that $\mathcal{Z}$ is non-empty and that (2.1) has a finite optimal solution. The optimization problem (2.1) is a fairly general formulation covering many important optimization problems. In fact, the linear objective function is the most general one, since any optimization problem can be expressed as one with a linear objective by considering its epigraph formulation, see e.g., Boyd and Vandenberghe (2004). [2] We will later on present a variety of relevant optimization and decision problems that can be represented in the general form (2.1).

As it is presented here, the general problem formulation (2.1) imposes only few assumptions on the sets $\mathcal{Z}_i$. In particular, it is not required that the sets $\mathcal{Z}_i$ are given in a certain explicit (computationally advantageous) form, and a processor might know the set $\mathcal{Z}_i$ only implicitly. However, in order to obtain a computational procedure, we need to impose an assumption on the constraint sets. The assumption we impose here is one of the least restrictive assumptions possible, i.e., we only require the existence of a "separation oracle" or "cutting-plane oracle".

Let us briefly discuss the conceptual idea of a cutting-plane oracle. We work in the following with half-spaces of the form $h := \{z : a^\top z - b \leq 0\}$, where $a \in \mathbb{R}^d$ and $b \in \mathbb{R}$. A half-space is called a *cutting-plane* if it satisfies the following properties.

Definition 2.7 (Cutting-Plane). *Let $\mathcal{S} \subset \mathbb{R}^d$ be a closed convex set and $z_q \notin \mathcal{S}$ a query point, a hyperplane $h(z_q)$ is called a cutting-plane that separates z_q from $\mathcal{S}$ if $a(z_q) \neq 0$ and*

$$a^\top(z_q) z \leq b(z_q) \quad \text{for all } z \in \mathcal{S}, \quad \text{and} \quad a^\top(z_q) z_q - b(z_q) = s(z_q) > 0. \tag{2.2}$$

[2] See also (A.3) in Appendix 1.

It is a consequence of the separating hyperplane theorem (i.e., Theorem A.3) that for every $z_q \notin \mathcal{S}$ there exists always at least one cutting-plane separating z_q and $\mathcal{S}$ strongly. In general, there exists an infinite number of possible cutting-planes that separate z_q and $\mathcal{S}$. As we are only interested in finding one cutting-plane for a given query point, we introduce the cutting-plane oracle as an algorithmic primitive.

Cutting-Plane Oracle $\mathrm{ORC}(z_q, \mathcal{S})$: Given a query point $z_q \in \mathbb{R}^d$ return

1. an empty h if $z_q \in \mathcal{S}$;
2. cutting-plane $h(z_q)$, separating z_q and $\mathcal{S}$, otherwise.

The cutting-plane oracle will be the key component for the distributed optimization algorithms we present in the following. However, in order to cope with the general problem formulation (2.1), we need to impose an additional, little restrictive, assumption on the cutting-plane oracle. The assumption follows from the general cutting-plane framework of Curtis Eaves and Zangwill (1971).

Assumption 2.1. *The cutting-planes generated by* $\mathrm{ORC}(z_q, \mathcal{S})$ *are such that (i) there exists a finite* L *such that* $\|a(z_q)\|_2 < L$ *and (ii)* $z_q(t) \to \bar{z}$ *and* $s(z_q(t)) \to 0$ *implies that* $\bar{z} \in \mathcal{S}$.

This assumption is not very restrictive and holds for many problem formulations, as we will discuss later on for several important problem classes.

Given a collection of cutting-planes $H = \bigcup_{k=1}^{m} h_k$, the polyhedron induced by these cutting-planes is $\mathcal{H} = \{z : A_H^\top z \leq b_H\}$, with the matrix $A_H \in \mathbb{R}^{d \times m}$ as $A_H = [a_1, \ldots, a_m]$, and the vector $b_H = [b_1, \ldots, b_m]^\top$.

Remark 2.8. *We refer to both a half-space* h *and the data inducing the half-space with a small italic letter. A collection of cutting-planes is denoted with italic capital letters, e.g.,* $H = \bigcup_{k=1}^{m} h_k$. *For a collection of cutting-planes, we denote the induced polyhedron with capital calligraphic letters, e.g.,* $\mathcal{H}$. *Please note the following notational aspect. A collection of cutting-planes* B *that is a subset of the cutting-planes contained in* H *is denoted as* $B \subset H$, *while the induced polyhedra satisfy* $\mathcal{B} \supseteq \mathcal{H}$.

Let now H be a collection of cutting-planes, with all cutting-planes $h_i \in H$ being generated as separating hyperplanes for some set $\mathcal{Z}_j$. Then the general optimization problem (2.1) is approximated by the linear *approximate program*

$$\begin{aligned} \text{maximize} \quad & c^\top z \\ \text{subj. to} \quad & A_H^\top z \leq b_H. \end{aligned} \tag{2.3}$$

In fact, the polyhedron $\mathcal{H} = \{z : A_H^\top z \leq b_H\}$ is an *outer approximation* of the original constraint set $\mathcal{Z} = \bigcap_{i=1}^{n} \mathcal{Z}_i$. It is easily seen that the optimal value of (2.3) is always an upper bound for the value of (2.1).

The conceptual idea we will pursue later on is to solve the general convex optimization problem (2.1) by solving a sequence of linear approximate problems (2.3). Obviously, solving a linear program is significantly easier than solving a general convex optimization problem. However, in the context of distributed optimization, we need to consider special solutions to the approximate linear programs. In particular, we will have to handle the non-uniqueness of optimal solutions in linear programming. We discuss next methods for linear programming with unique solutions.

2.3.2 Unique Solution Linear Programming

We review in this part some important but partially non-standard concepts from linear programming. The key focus will be in the following on the uniqueness of the optimal solutions. We focus here on linear programs in the *primal form*

$$\begin{aligned} \text{maximize} \quad & c^\top z \\ \text{subj. to} \quad & A_H^\top z \leq b_H. \end{aligned} \tag{2.4}$$

Sometimes, we will also refer also to the Lagrange dual of (2.4)

$$\begin{aligned} \text{minimize} \quad & b_H^\top y \\ \text{subj. to} \quad & A_H y = c, \quad y \geq 0. \end{aligned} \tag{2.5}$$

An optimal solution y to (2.5) will be the Lagrange multiplier of the primal problem (2.4).

In what follows, we will denote the optimal value of (2.4) by γ_H, i.e., $\gamma_H := \max_{z \in \mathcal{H}} c^\top z$, where $\mathcal{H} = \{z : A_H^\top z \leq b_H\}$. It is a standard results in linear programming that strong duality holds and thus γ_H is also the optimal value of (2.5). The linear program (2.4) has in general several optimizers, and we denote the set of all (primal) optimizers of (2.4) with

$$\Gamma_H := \{z \in \mathcal{H} : c^\top z \geq c^\top v, \forall v \in \mathcal{H}\}. \tag{2.6}$$

Clearly, Γ_H is always a polyhedral set. If the set Γ_H contains more than a single point, we say that the linear program (2.4) is degenerate. Otherwise, we say it is non-degenerate. For a discussion on degeneracy of linear programs in the dual form (2.5), we refer to Bürger et al. (2012b) or Jones et al. (2007). Standard linear programming algorithms, such a simplex or interior point methods, can efficiently compute an optimal solution to the linear program. However, it depends on a variety of aspects, such as initial conditions, step-size or pivot rules, which solution in Γ_H will be computed. We are interested here in finding one particular and predictable solution to the linear program. Therefore, we cannot simply apply standard solution algorithms, but need to specify additional criteria for selecting solutions from Γ_H. Naturally, the chosen criteria have to be such that the optimal solution is also efficiently computable. We discuss next two selection criteria for optimal solutions, which turn out to be computationally attractive. We want to point out that the role of unique solutions to linear programs and the different selection criteria were discussed already in Amenta (1993).

Lexicographic Minimal Solution The first selection criterion we discuss is the *lexicographic* selection rule. Lexicographic rules allow us to compare vectors. First, the notion of lex-positivity is needed.

Definition 2.9 (Lex-positivity). *A vector $v = (v_1, \dots, v_r)$ is said to be* lexico-positive *(or lex-positive) if $v \neq 0$ and the first non-zero component of v is positive.*

Lex-positivity allows us to define a unique ordering of vectors. In particular, we can say that $v \succ u$ if $v - u$ is lex positive. Now, a natural criterion to select a unique optimal solution of a linear program, is simply to take the lexicographically minimal vector in the set Γ_H, i.e., the unique vector $z \in \Gamma_H$ such that $v - z$ is lex-positive for all $v \in \Gamma_H$.

The idea of the lexicographic minimal solution is illustrated in Figure 2.2. The feasible region is defined by two linear constraints, where one constraint is orthogonal to the cost vector c. Therefore the optimal solution is not unique. However, the point L indicates the unique lexicographic minimal solution. In fact, any point on the horizontal constraint which is to the right of L is an optimal solution, but L is minimal in the horizontal component. Now, an important property of the lexicographic optimal solution is that it can be directly associated to a family of "perturbed" objective functions.

Proposition 2.10. *Let a set of half-spaces define the polyhedron $\mathcal{H}$ and let z_H^* be the lexicographic minimal solution to* (2.4). *Consider the linearly perturbed objective function*

$$J_\epsilon = c^\top z - \boldsymbol{\epsilon}^\top z,$$

with $\boldsymbol{\epsilon} = [\epsilon, \epsilon^2, \ldots, \epsilon^d]$ for some $\epsilon > 0$. Then there exists $\bar{\epsilon} > 0$ such that for any $\epsilon \in (0, \bar{\epsilon}]$

$$z_H^* = \arg\max_{z \in \mathcal{H}} J_\epsilon(z). \tag{2.7}$$

This is a classical result, proven for example in Murty (1983), and we outline here only the proof idea. Consider the polynomial $p(\epsilon) := \epsilon z_1 + \epsilon^2 z_2 + \ldots + \epsilon^d z_d$. In Dantzig (1963) Section 10.2, Lemma 1, is shown that there exists $\bar{\epsilon}$, such that for all $\epsilon \in [0, \bar{\epsilon}]$ $p(\epsilon) > 0$ if and only if $z = [z_1, \ldots, z_d] \succ 0$. Consider now the linear program

$$\min \boldsymbol{\epsilon}^\top z, \; c^\top z \geq \gamma_H, A_H^\top z \leq b. \tag{2.8}$$

with $\boldsymbol{\epsilon} = [\epsilon, \epsilon^2, \ldots, \epsilon^d]$. For any ϵ sufficiently small, an optimal point of (2.8) will be the lexicographic minimum of the original linear program. Furthermore, one can construct, following for example the classical ideas form Mangasarian and Meyer (1979), from the primal and dual solution of (2.8) a KKT point of the perturbed problem (2.7).

The most important property of the lexicographically perturbed objective function, we will exploit here, is as follows. Let $\{\hat{z}(k)\}$ and $\{\tilde{z}(k)\}$ be two sequences satisfying $\lim_{k \to \infty} |J_\epsilon(\hat{z}(k)) - J_\epsilon(\tilde{z}(k))| \to 0$ for some $\epsilon > 0$, then $\lim_{k \to \infty} \|\hat{z}(k) - \tilde{z}(k)\| \to 0$.

The use of lexicographic minimal solutions in distributed linear programming has been proposed in Notarstefano and Bullo (2011). We have shown in the context of distributed linear programming that the lexicographic minimum can be computed using a refined version of the Simplex algorithm as shown in Bürger et al. (2012b), see also Bürger et al. (2011b), Bürger et al. (2011a). In fact, lexicographic minimal solutions can be computed in by a *Simplex algorithm* that uses a certain pivot rule. We refer here to Bürger et al. (2012b) and the references therein for a detailed discussion on the use of lexicographic optimal solutions for distributed linear programs. Lexicographic optimal solutions are very intuitive unique solution rules in a linear programming framework. However, they come along with some disadvantages. In fact, although a variant of the Simplex algorithm can be used to compute the lex-minimal solution, this algorithm is not necessarily efficient. A restriction to this algorithm prohibits the use of modern and more efficient optimization algorithms. A second

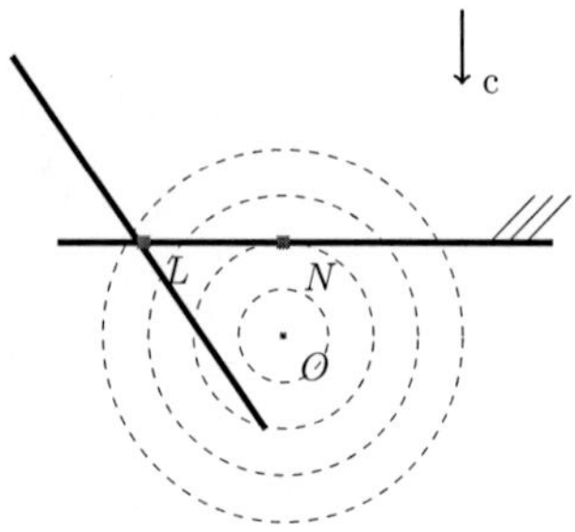

Figure 2.2: Different solution concepts for degenerate linear programs. The lexicographic optimal solution refers to the point L, while the unique solution with minimal 2-norm is shown as the point N. All points on the horizontal constraint on the right side of L are optimal solutions.

disadvantage is that the lexicographic minimal solution can be unbounded, although the linear program has a finite optimal solution. To see this, consider the situation illustrated in Figure 2.2. Suppose now that the orientation of the horizontal coordinate is reversed, then the lexicographic optimal solution would be unbounded. Algorithms computing lexicographic optimal solutions might therefore eventually falsely detect unboundedness of the problem.

Minimal 2-Norm Solution An alternative solution concept for linear programming, overcoming the drawbacks of lexicographic minimal solutions is the minimal 2-norm solution. The minimal 2-norm solution to a linear program is defined as

$$z_H^* = \arg \min_{z \in \Gamma_H} \|z\|_2. \tag{2.9}$$

Note that we seek here the element with minimal two norm in the set of optimal solutions Γ_H, and not in the set of feasible solutions. The idea of the minimal 2-norm solution is also illustrated in Figure 2.2. The optimal solution, having minimal distance to the origin O is clearly the point N. Note that the minimal 2-norm solution (N) is in general different from the lexicographic minimal solution (L). A first observation we can make is that the minimal 2-norm solution is always finite, if the linear program (2.4) has a finite optimal solution. Thus, the 2-norm solution overcomes one drawback of the lexicographic solution.

A second advantage is that the 2-norm solution can be efficiently computed. In fact, finding a minimal norm solution to a linear program is a classical problem. Starting from the early reference Mangasarian (1983) research on this topic is still actively pursued Zhao and Li (2002). The main observation we make here is that the minimal 2-norm solution to a linear program can be computed as the solution of a quadratic program, as we show in the next result.

Proposition 2.11. *Let $u^* \in \mathbb{R}^{|H|}, \alpha^* \in \mathbb{R}, l^* \in \mathbb{R}^d$ be the optimal solution to*

$$\begin{aligned} \min_{u,\alpha,l} \quad & \frac{1}{2}(A_H u + \alpha c)^\top (A_H u + \alpha c) + b_H^\top u + c^\top l \\ \text{s.t.} \quad & A_H^\top l - \alpha b \geq 0, \ u \geq 0 \end{aligned} \tag{2.10}$$

then $z_H^* = -A_H u^* - \alpha^* c$ solves (2.9).

Proof. The minimal 2-norm solution is the solution to

$$\min_{z,y} \frac{1}{2} z^\top z, \text{ s.t. } A_H^\top z \leq b_H, \ A_H y = c, \ c^\top z - b_H^\top y = 0, \ y \geq 0, \tag{2.11}$$

where the constraints represent the linear programming optimality conditions (KKT-conditions). The Lagrangian of (2.11) can be directly determined to be

$$\mathcal{L}(z, y, u, l, \alpha) = \frac{1}{2} z^\top z + u^\top (A_H^\top z - b_H) + l^\top (A_H y - c) + \alpha (c^\top z - b_H^\top y), \ y, u \geq 0. \tag{2.12}$$

It follows now that $y^* = \arg\min_{y \geq 0} \mathcal{L}(z, y, u, l, \alpha) = 0$ if $A_H^\top l - \alpha b_H \geq 0$. From

$$z^* = \arg\min_z \ \mathcal{L}(z, y, u, l, \alpha)$$

follows that $z^* = -A_H u - \alpha c$. Consequently, the problem (2.10) as stated in the proposition is now

$$\min_{u \geq 0, l, \alpha} \ -\mathcal{L}(z^*, y^*, u, l, \alpha).$$

□

The 2-norm solution to a linear program can therefore be computed by any optimization algorithm that can solve convex quadratic programs. Thus, all the computational advances on optimization algorithms can be exploited. Aside from the computational issue, the minimal 2-norm solution has an additional, very desirable property. As we will show next, the minimal 2-norm solution always maximizes a strongly concave cost function.

Lemma 2.12. *Let a set of cutting-planes define the polyhedron $\mathcal{H}$ and let z_H^* be the minimal 2-norm solution to* (2.4). *Consider the quadratically perturbed linear objective*

$$J_\epsilon(z) = c^\top z - \frac{\epsilon}{2} \|z\|_2^2$$

parametrized with a constant $\epsilon > 0$. Then there exists a $\bar{\epsilon} > 0$ such that for any $\epsilon \in [0, \bar{\epsilon}]$

$$z_H^* = \arg\max_{z \in \mathcal{H}} \ J_\epsilon(z). \tag{2.13}$$

This result is the counterpart to Proposition 2.10. The proof of this result is similar to the classical proof presented in Mangasarian and Meyer (1979). However, since the considered set-up is slightly different and the result is fundamental for the methodologies developed here, we present the proof.

Proof. The minimal 2-norm solution z_H^* is the unique minimizer of

$$\min_z \frac{1}{2}\|z\|^2, \quad \text{s.t. } c^\top z \geq \gamma_H,\ A_H^\top z \leq b_H.$$

and satisfies therefore the feasibility conditions $c^\top z_H^* = \gamma_H$ and $A_H^\top z_H^* \leq b_H$. Since z_H^* is an optimal solution, there exist multipliers $\mu^* \in \mathbb{R}$ and $\lambda^* \in \mathbb{R}_{\geq 0}^{|H|}$, such that the KKT conditions are satisfied, i.e.,

$$z_H^* - \mu^* c + A_H \lambda^* = 0 \tag{2.14}$$

$$\lambda^{*\top} A_H^\top z_H^* - \lambda^{*\top} b_H = 0. \tag{2.15}$$

Since z_H^* is also a solution to the original linear program (2.3), there also exists a multiplier vector $y^* \in \mathbb{R}_{\geq 0}^{|H|}$ satisfying the linear programming optimality conditions

$$-c + A_H y^* = 0 \tag{2.16}$$

$$y^{*\top} A_H^\top z_H^* - b_H^\top y^* = 0. \tag{2.17}$$

We have to show now that the existence of z_H^*, μ^*, λ^* and y^* imply, for a sufficiently small ϵ, the existence of a multiplier vector π^* satisfying the optimality conditions of (2.13), which are

$$-c + \epsilon z_H^* + A_H \pi^* = 0 \quad \text{and} \quad \pi^{*\top} A_H^\top z_H^* - \pi^* b_H = 0. \tag{2.18}$$

We distinguish now the two cases $\mu^* > 0$ and $\mu^* = 0$. First, assume $\mu^* > 0$. We can multiply (2.16) with $\frac{t}{\mu^*}$, for arbitrary $t \in (0, 1]$, and add to this (2.16), multiplied by $(1-t)$ to obtain

$$\frac{t}{\mu^*} z_H^* - c + A_H\left(\frac{t}{\mu^*}\lambda^* + (1-t)y^*\right) = 0. \tag{2.19}$$

The same steps can be repeated with (2.15) and (2.17) to obtain

$$\left(\frac{t}{\mu^*}\lambda^{*\top} + (1-t)y^{*\top}\right)(A_H^\top z_H^* - b_H) = 0. \tag{2.20}$$

With (2.19) and (2.20), for any $\epsilon \leq \frac{1}{\mu^*}$, one can define $t_\epsilon = \epsilon\mu^*$. Then $\pi^* = \frac{t_\epsilon}{\mu^*}\lambda^{*T} + (1-t_\epsilon)y^{*T}$ solves (2.18). In the second case $\mu^* = 0$, one can pick an arbitrary $\epsilon > 0$, multiply (2.14) (and (2.15), respectively) with ϵ and add (2.16) (or (2.17), respectively) to obtain $\epsilon z_H^* - c + A_H(\epsilon\lambda^* + y^*) = 0$ and $(\epsilon\lambda^* + y^*)(A_H^\top z_H^* - b_H) = 0$. Now, $\pi^* := (\epsilon\lambda^* + y^*)$ solves (2.18). □

Since the perturbed cost functions are strictly concave, we also have, as for the lexicographically perturbed cost, the following property. Let $\{\hat{z}(k)\}$ and $\{\tilde{z}(k)\}$ be two sequences satisfying $\lim_{k\to\infty} |J_\epsilon(\hat{z}(k)) - J_\epsilon(\tilde{z}(k))| \to 0$ for some $\epsilon > 0$, then $\lim_{k\to\infty} \|\hat{z}(k) - \tilde{z}(k)\| \to 0$.

The algorithm and the proofs, we present later on, will work with both unique solution rules. The important property of the two solution concepts is that over any constraint set there is a *unique* optimum. If not explicitly stated differently, we will in the following only refer to a *unique solution rule*, and will not distinguish whether the lexicographic optimum or the minimal 2-norm solution is considered. In this spirit, we will also use J_ϵ to refer to either of the two perturbed objective functions.

Basis and Combinatorial Dimension Before we can proceed to the presentation of a general distributed optimization algorithm, we first have to introduce the concept of a basis, see e.g., Dyer et al. (2004). We assume in the following that H is a collection of cutting-planes.

Definition 2.13 (Basis). *A subset $B \subseteq H$ is a basis of H if the solution to the linear program defined with the constraint set B, say z_B^*, is identical to solution of the linear program defined with the constraint set H, say z_H^*, i.e., $z_B^* = z_H^*$, and for any strict subset of cutting-planes $B' \subset B$, it holds that $z_{B'}^* \neq z_B^*$.*

We will later on also refer to a basis of the original problem (2.1), as a basis of some cutting-planes defined to approximate (2.1). Note that the definition of a basis depends on the choice of the objective function. Consider again the situation illustrated in Figure 2.2. If the lexicographically minimal solution (L) is considered, the two illustrated constraints define together a basis. In fact, considering the lexicographic rule, a basis will always correspond to a vertex of the feasible polytope. However, if the minimal 2-norm solution (N) is considered, already the horizontal constraint defines the basis. Therefore, a basis for the 2-norm solution does not necessarily correspond to a vertex of the feasible set. Now, as we have seen that the number of cutting-planes defining a basis is not fixed, we might consider the maximal number of constraints required.

Definition 2.14 (Combinatorial Dimension). *The combinatorial dimension d is the maximum number of cutting-planes necessary to define a basis.*

It is a standard result in linear programming that any solution to a (feasible) linear program of the form (2.3) is fully determined by at most d constraints. That is, the combinatorial dimension is $\delta = d$.

Remark 2.15. *The definition of a basis that we employ here follows the one considered in "abstract programming" or "generalized linear programming" (Amenta, 1993), (Gärtner and Welzl, 1996b), (Dyer et al., 2004). In more classical references on linear programming, a basis is often defined as a set of constraints H such that the matrix A_H is invertible. With this definition, a basis corresponds always to a vertex of the polyhedral constraint set. A fundamental result in linear programming states that any feasible linear program has an optimal solution at a vertex of the constraint set (Luenberger, 1973). Defining the basis in terms of the matrix A_H is advantageous when working with simplex algorithms, that move along the vertices of the constraint set. However, in the context of this work the given definition of a basis is clearly preferable.*

We assume in this definition implicitly that the linear program (2.4) is solved using one of the selection rules presented above, such that the optimal solution is always unique. It is not always a trivial task to compute a basis for a linear program. However, in most cases the following observation proves to be useful.

Proposition 2.16. *The basis for a linear program (2.4) with optimal solution z_H^* is always a subset of the active constraints at z_H^*. If the problem is non-degenerate, the basis is identical to the set of active constraints.*

In fact, very often it will be sufficient to find the active constraints in order to compute a basis. The active constraints can naturally be easily detected. With these results, we are ready to present in the next section a fully distributed, asynchronous algorithm, which solves general optimization problems of the form (2.1).

2.3.3 The Algorithm Definition

We propose in this section a general framework for distributed optimization based on the cutting-plane concept. The main feature of the novel algorithm are its little requirements on both, the problem structure and the communication network. We will focus in this section on the presentation and the discussion of the general algorithm, and discuss several specification for more precise problem formulations later on. The algorithm to solve general distributed optimization problems (2.1) is as follows.

Cutting-Plane Consensus: Processors store and update collections of cutting-planes. The cutting-planes stored by agent $i \in \mathbf{V}$ at iteration t are always a basis of a corresponding linear approximate program (2.3), and are denoted by $B^{[i]}(t)$. A processor initializes its local collection of cutting-planes $B_0^{[i]}$ with a set of cutting-planes chosen such that $\mathcal{B}_0^{[i]} \supset \mathcal{Z}_i$ and $\max_{z \in \mathcal{B}_0^{[i]}} c^\top z < \infty$.

Each processor repeats then the following steps:

(S1) it transmits its current basis $B^{[i]}(t)$ to all its out-neighbors $\mathcal{N}_O(i,t)$ and receives the basis of its in-neighbors $Y^{[i]}(t) = \bigcup_{j \in \mathcal{N}_I(i,t)} B^{[j]}(t)$;

(S2) it defines $H_{tmp}^{[i]}(t) = B^{[i]}(t) \cup Y^{[i]}(t)$, and computes (i) a query point $z^{[i]}(t)$ as a *unique solution to the approximate program* (2.3), and (ii) a minimal set of active constraints $B_{tmp}^{[i]}(t)$;

(S3) it calls the cutting-plane oracle for the constraint set $\mathcal{Z}_i$ at the query point $z^{[i]}(t)$ and eventually generates a new cutting-plane;

(S4) it updates its collection of cutting-planes as a basis of the constraint set defined by $B_{tmp}^{[i]}(t)$ and the eventually generated cutting-plane.

The four steps of the algorithm can be summarized as *communication* (S1), *computation* of the query point (S2), *generation* of a new cutting-plane (S3) and *dropping* of all inactive constraints (S4). A more formal description of the algorithm is presented in Algorithm 1, where we use the formalism introduced in Section 2.2. Note that we denote the "artificial" constraints used for the initialization of the algorithm with B_M. In step (S2) of the algorithm a unique solution rule is applied to solve the linear program (2.3). The algorithm works with both, the lexicographic minimal solution and the minimal 2-norm solution. Naturally, the selected unique solution rule should not be changed during the evolution of the algorithm. In practice, the minimal 2-norm solution has some computational advantages and seems preferable. Please note again that the only assumption we made on the underlying optimization problem is the existence of a suitable cutting-plane oracle. In fact, only step (S3) depends on the actual representation of the optimization problem, while all other steps are independent of the problem representation. Thus, the cutting-plane consensus algorithm provides a conceptual framework for solving various distributed optimization problems.

Algorithm 1 cutting-plane consensus

`Processor state:`	$B^{[i]}(t)$
`Initialization:`	$B^{[i]}(0) := B_M$
`Function MSG:`	**Return** $B^{[i]}(t)$.
`Function STF:`	**Set** $Y^{[i]}(t) := \cup_{j\in\mathcal{N}_I(i,t)}\texttt{MSG}(B^{[j]}(t))$; (S1) $H^{[i]}_{tmp}(t) \leftarrow B^{[i]}(t) \cup Y^{[i]}(t)$; (S2) $z^{[i]}(t) \leftarrow$ unique solution to $\min_{z\in\mathcal{H}^{[i]}_{tmp}(t)} c^\top z$; $B^{[i]}_{tmp}(t) \leftarrow$ set of active constraints at $z^{[i]}(t)$; (S3) $h(z^{[i]}(t)) \leftarrow \mathrm{ORC}(z^{[i]}(t), \mathcal{Z}_i)$; (S4) $B^{[i]}(t+1) \leftarrow$ Basis of $B^{[i]}_{tmp}(t) \cup h(z^{[i]}(t))$.

The cutting-plane consensus algorithm is explicitly designed for the use in processor networks, and we want to emphasize here four important aspects of the algorithm.

Distributed Initialization: Each processor can initialize the local constraint sets as a basis of the artificial constraint set $\{z \in \mathbb{R}^d : -M\mathbf{1} \leq z \leq M\mathbf{1}\}$ for some $M \gg 1$. If $M \in \mathbb{R}_{>0}$ is chosen sufficiently large, the artificial constraints will be dropped during the evolution of the algorithm.

Bounded Communication: Each processor stores and transmits at most $(d+1)d$ numbers at a time. In particular, processors exchange bases of (2.3), which are defined by not more than d cutting-planes. Each cutting-plane is fully defined by $d+1$ numbers.

Bounded Local Computations: Each processor has to compute locally the solution to a linear program with $d(|\mathcal{N}_I(i,t)|+1)$ constraints according to a unique solution rule.

Asynchronous Communication: The cutting-plane consensus algorithm does not require a time-synchronization. Each processor can perform its local computations at any speed and update its local state whenever it receives data from some of its in-neighbors.

Due to these properties, the cutting-plane consensus algorithm is particularly well suited for optimization in large-scale processor networks with unreliable communication. Based on these observations, we can conclude that the *space-complexity* (i.e., the number of bytes stored by one processor) is in $\mathcal{O}(d^2 \cdot |\mathcal{N}_I|)$, while the *communication complexity* is in $\mathcal{O}(d^2)$. It seems remarkable that neither the space nor the communication complexity of the cutting-plane consensus algorithm depends on the number of processors in the network.

2.3.4 Technical Analysis

We want to discuss now the technical aspects of the cutting-plane consensus algorithm. Before presenting the proof of the algorithm correctness, we point out three important technical properties related to its evolution:

- The linear constraints stored by a processor form always a *polyhedral outer approximation* of the globally feasible set $\mathcal{Z}$.
- The cost-function of each processor is monotonically non-increasing over the evolution of the algorithm.
- If the communication graph $\mathcal{G}_c$ is a strongly connected *static* graph, then after $\mathrm{diam}(\mathcal{G}_c)$ communication rounds, all processors in the network compute a query-point with a cost not worse than the best processor at the initial iteration.

These properties provide an intuition about the functionality of the algorithm and the line we will follow to prove its correctness. They are formalized and proven rigorously in the following result.

Lemma 2.17. *Let $z^{[i]}(t)$ be the query point and $B^{[i]}(t)$ the corresponding basis. Let $\mathcal{B}^{[i]}(t) \subset \mathbb{R}^d$ be the feasible set induced by $B^{[i]}(t)$. Then,*

(i) $\mathcal{B}^{[i]}(t) \supset \mathcal{Z}$ *for all* $i \in \{1 \ldots, n\}$ *and* $t \geq 0$*;*

(ii) $\lim_{t\to\infty} z^{[i]}(t) = \bar{z}$ *and* $\bar{z} \in \mathcal{Z}$ *implies* $\bar{z}$ *is a minimizer of* (2.1)*;*

(iii) for any of the two unique solution rules, there exists $\underline{\epsilon} > 0$ *such that for all* $i \in \{1, \ldots, n\}$ *and all* $t \geq 0$*, the query points* $z^{[i]}(t)$ *maximize the objective function* $J_\epsilon(z)$ *over the set of constraints* $B^{[i]}(t) \cup Y^{[i]}(t)$ *(as defined in (S2)) for all* $\epsilon \in [0, \underline{\epsilon}]$*;*

(iv) $J_{\underline{\epsilon}}(z^{[i]}(t+1)) \leq J_{\underline{\epsilon}}(z^{[i]}(t))$ *for all* $i \in \{1, \ldots, n\}$ *and all* $t \geq 0$*;*

(v) if $\mathcal{G}_c$ *is a strongly connected* static *graph, then* $J_{\underline{\epsilon}}(z^{[j]}(t + \mathrm{diam}(\mathcal{G}_c))) \leq J_{\underline{\epsilon}}(z^{[i]}(t))$ *for all* $i, j \in \{1, \ldots, n\}$ *and all* $t \geq 0$.

Proof. To see (i), note that any cut h_k generated by the oracle of processor i, $\mathrm{ORC}(\cdot, \mathcal{Z}_i)$ is such that the half-space h_k contains $\mathcal{Z}_i$, and in particular h_k contains $\mathcal{Z} = \bigcap_{i=1}^n \mathcal{Z}_i$. Thus any collection of cuts $H = \bigcup_k h_k$, generated by arbitrary processors is such that $\mathcal{H} \supset \bigcap_{i=1}^n \mathcal{Z}_i = \mathcal{Z}$, and in particular $\mathcal{B}^{[i]}(t) \supset \mathcal{Z}$. The claim (ii) follows since $z^{[i]}(t)$ is computed as a maximizer of the linear cost $c^\top z$ over the collection of cutting-planes $H^{[i]}_{tmp}(t)$. The induced polyhedron is such that $\mathcal{H}^{[i]}_{tmp}(t) \supset \mathcal{Z}$. Therefore, we can conclude that $c^\top z^{[i]}(t) \geq c^\top z^*$, where z^* is an optimizer of (2.1). By continuity of the linear objective function, we have that $c^\top \bar{z} \geq c^\top z^*$ On the other hand, $c^\top z \leq c^\top z^*$ for all $z \in \mathcal{Z}$. This proves the statement. The statement (iii) follows from Lemma 2.12. For any approximate program defined by processor i at time t, there exists a constant $\bar{\epsilon}_{it} > 0$ such that $z^{[i]}(t)$ is the unique maximizer of the family of either the lexicographically perturbed cost $J_\epsilon = c^\top z - [\epsilon, \epsilon^2, \ldots, \epsilon^d]z$, or strictly concave objective functions $J_\epsilon(z) := c^\top z - \frac{\epsilon}{2}\|z\|^2, \quad \epsilon \in [0, \bar{\epsilon}_{it}]$, over the set of constraints $B^{[i]}(t) \cup Y^{[i]}(t)$. One can now always find $\underline{\epsilon} > 0$ such that $\underline{\epsilon} \leq \bar{\epsilon}_{it}$ for all $i \in \{1, \ldots, n\}$ and $t \geq 0$. To see claim (iv), note that adding cutting-planes, either by receiving them from neighbors (S2) or by generating them with the oracle (S3), can only decrease the value of the strictly concave objective function $J_{\underline{\epsilon}}(\cdot)$ and the basis computation in (S4) keeps, by its definition, the value of $J_{\underline{\epsilon}}(\cdot)$ constant. Finally, (v) can be seen as follows. Starting at any time t at some processor i, at time $t+1$ all processors in $l \in \mathcal{N}_I(i, t)$ received the basis of processor i, and compute a query point that satisfies $J_{\underline{\epsilon}}(z^{[l]}(t+1)) \leq J_{\underline{\epsilon}}(z^{[i]}(t))$ for all $l \in \mathcal{N}_I(i, t)$. This argument can be repeatedly applied to see that, in the static, strongly connected communication graph $\mathcal{G}_c$, at least after diam$(\mathcal{G}_c)$ iterations, all processors in the network have an objective value smaller than $J_{\underline{\epsilon}}(z^{[i]}(t))$. $\square$

We are now ready to proof the correctness of the algorithm. Please note that the following proofs will rely strongly on the use of the parameterized cost function $J_{\underline{\epsilon}}(\cdot)$, that can either be the lexicographically or the quadradically perturbed cost. However, for the clarity of presentation we will simplify our notation in the following proofs and write simply $J(\cdot)$ instead of $J_{\underline{\epsilon}}(\cdot)$. We start by formalizing two auxiliary results which are also interesting on their own. The first result states the convergence of the query points to the locally feasible sets.

Lemma 2.18 (Convergence). *Assume Assumption 2.1 holds. Let $z^{[i]}(t)$ be the query point generated by processor i performing the cutting-plane consensus algorithm. Then, the sequence $\{z^{[i]}(t)\}_{t\geq 0}$ has a limit point in the set $\mathcal{Z}_i$, i.e., there exists $\bar{z}^{[i]} \in \mathcal{Z}_i$ such that*

$$\lim_{t\to\infty} \|z^{[i]}(t) - \bar{z}^{[i]}\|_2 \to 0.$$

Proof. All $z^{[i]}(t)$ are computed as maximizers of the common strictly concave objective function $J(\cdot)$ (Lemma 2.17 (iii)) and $J(\cdot)$ is monotonically non-increasing over the sequence of query points computed by a processor (Lemma 2.17 (iv)). Any sequence $\{J(z^{[i]}(t))\}_{t\geq 0}$, $i \in \{1, \ldots, n\}$, has therefore a limit point, i.e., $\lim_{t\to\infty} J(z^{[i]}(t)) \to \bar{J}^{[i]}$. Since the sequence is convergent, it holds that $\lim_{t\to\infty} \big(J(z^{[i]}(t)) - J(z^{[i]}(t+1))\big) \to 0$.

By the properties of the unique minimum objective functions, follows that $\lim_{t\to\infty} \|z^{[i]}(t) - z^{[i]}(t+1)\|_2 \to 0$ and the sequence of query points has a limit point, i.e., $\lim_{t\to\infty} \|z^{[i]}(t) - \bar{z}^{[i]}\|_2 \to 0$. Suppose now, to get a contradiction, that $\bar{z}^{[i]} \notin \mathcal{Z}_i$. Then there exists $\delta > 0$ such that all z satisfying $\|z - z^{[i]}\|_2 < \delta$ are not contained in $\mathcal{Z}_i$. Since $\lim_{t\to\infty} \|z^{[i]}(t) - \bar{z}^{[i]}\|_2 \to 0$, there exists a time instant T_δ such that $\|z^{[i]}(t) - \bar{z}^{[i]}\|_2 < \delta$ for all $t \geq T_\delta$, and thus $z^{[i]}(t) \notin \mathcal{Z}_i$ for $t \geq T_\delta$. But now, for all $t \geq T_\delta$ the oracle ORC$(z^{[i]}(t), \mathcal{Z}_i)$ will generate a cutting-plane according to (2.2), cutting off $z^{[i]}(t)$. According to (2.2), it must hold that $a^\top(z^{[i]}(t))z^{[i]}(t) - b(z^{[i]}) = s(z^{[i]}(t)) > 0$ and $a^\top(z^{[i]}(t))z^{[i]}(t+1) - b(z^{[i]}) \leq 0$. This implies that $a^\top(z^{[i]}(t))\big(z^{[i]}(t) - z^{[i]}(t+1)\big) \geq s(z^{[i]}(t))$ and consequently $\|z^{[i]}(t) - z^{[i]}(t+1)\|_2 \geq (\|a(z^{[i]}(t))\|_2)^{-1} s(z^{[i]}(t))$. By Assumption 2.1 (i) holds $\|a(z^{[i]}(t))\|_2 < \infty$ and thus $\lim_{t\to\infty} s(z^{[i]}(t)) \to 0$. As a consequence of Assumption 2.1 (ii) follows directly that $\bar{z}^{[i]} \in \mathcal{Z}_i$, providing the contradiction. $\square$

The second result shows that all processors in the network will reach an agreement.

Lemma 2.19 (Agreement). *Assume the communication network $\mathcal{G}_c(t)$ is jointly strongly connected. Let $z^{[i]}(t)$ be the query points generated by the cutting-plane consensus algorithm, then*

$$\lim_{t\to\infty} \|z^{[i]}(t) - z^{[j]}(t)\|_2 \to 0, \quad \textit{for all } i,j \in \{1,\ldots,n\}.$$

Proof. Let $\bar{J}^{[i]} := J(\bar{z}^{[i]})$ be the objective value of the limit point $\bar{z}^{[i]}$ of the sequence $\{z^{[i]}(t)\}_{t\geq 0}$ computed by processor i. We show first that the limiting objective values $\bar{J}^{[i]}$ are identical for all processors. Suppose by contradiction that there exist two processors, say i and j, such that $\bar{J}^{[i]} < \bar{J}^{[j]}$. Pick now $\delta_0 > 0$ such that $\bar{J}^{[j]} - \bar{J}^{[i]} > \delta_0$. The sequences $\{J(z^{[i]}(t))\}_{t\geq 0}$ and $\{J(z^{[j]}(t))\}_{t\geq 0}$ are monotonically increasing and convergent. Thus, for every $\delta > 0$ there exists a time T_δ such that for all $t \geq T_\delta$, $J(z^{[i]}(t)) - \bar{J}^{[i]} \leq \delta$ and $J(z^{[j]}(t)) - \bar{J}^{[j]} \leq \delta$. This implies that there exists T_{δ_0} such that for all $t \geq T_{\delta_0}$,

$$J(z^{[i]}(t)) \leq \delta_0 + \bar{J}^{[i]} < \bar{J}^{[j]}.$$

Additionally, since the objective functions are non-increasing, it follows that for any time instant $t' \geq 0$, $J(z^{[j]}(t')) \geq \bar{J}^{[j]}$. Thus, for all $t \geq T_{\delta_0}$ and all $t' \geq 0$,

$$J(z^{[i]}(t)) < J(z^{[j]}(t')). \tag{2.21}$$

Pick now $t_0 \geq T_{\delta_0}$. For all $\tau \geq 0$ define now an index set I_τ as follows: Set $I_0 = \{i\}$ and for any $\tau \geq 0$ define I_τ by adding to $I_{\tau-1}$ all indices k for which there exist some $l \in I_{\tau-1}$ such that $(k, l) \in E(t_0 + \tau)$. Since, by assumption $\mathcal{G}_c^\infty(t_0)$ is strongly connected, the set I_τ will eventually include all indices $1, \dots, N$, and in particular there is τ^* such that $j \in I_{\tau^*}$. The algorithm is such that for all $l \in I_\tau$, $J(z^{[l]}(t_0 + \tau)) \leq J(z^{[i]}(t_0))$ and thus

$$J(z^{[j]}(t_0 + \tau^*)) \leq J(z^{[i]}(t_0)). \tag{2.22}$$

But (2.22) contradicts (2.21), proving that $\bar{J}^{[i]} = \bar{J}^{[2]} = \dots = \bar{J}^{[n]} =: \bar{J}$. Thus, it must hold that for all $i, j \in \{1, \dots, N\}$, $\lim_{t\to\infty} |J(z^{[i]}(t)) - J(z^j(t))| \to 0$. From the properties of the perturbed objective functions follows now $\lim_{t\to\infty} \|z^{[i]}(t) - z^{[j]}(t)\|_2 \to 0$, which proves the theorem. □

The correctness of the algorithm is summarized in the following theorem.

Theorem 2.20 (Correctness). *Let $\mathcal{G}_c(t)$ be a jointly strongly connected communication network with processors performing the cutting-plane consensus algorithm, and let Assumption 2.1 hold. Let z^* be the unique optimizer to* (2.1) *according to the chosen unique optimum rule, then*

$$\lim_{t\to\infty} \|z^{[i]}(t) - z^*\|_2 \to 0 \quad \textit{for all } i \in \{1, \dots, n\}.$$

Proof. It follows from Lemma 2.19 that the query points of all processors converge to the same query point, i.e., $\bar{z}^{[i]} = \bar{z}$ for all processors i. Now, we can conclude from Lemma 2.18 that $\bar{z} \in \mathcal{Z}_i$ for all i and thus $\bar{z} \in \mathcal{Z}$. It follows now from Lemma 2.17, part (ii), that $\bar{z}$ is an optimal solution to (2.1). It remains to show that $\bar{z}$ is the optimal solution according to the chosen rule. Let z^* be the unique optimal solution. Then there exists an $\epsilon > 0$ such that the parameterized objective function satisfies $J_\epsilon(z^*) > J_\epsilon(z)$ for all $z \in \mathcal{Z}$ and $J_\epsilon(z^{[i]}(t)) \geq J_\epsilon(z^*)$ for all t. With the same argumentation used for Lemma 2.17, part (ii), we conclude that $\bar{z}$ is the unique solution maximizing $J_\epsilon(\cdot)$ over $\mathcal{Z}$, i.e., $\bar{z}$ is the corresponding unique optimal solution to (2.1). □

A major advantage for using the cutting-plane consensus algorithm in distributed systems is its inherent fault-tolerance. The requirements on the communication network are very weak and the algorithm can well handle disturbances in the communication like, e.g., packet-losses or delays. Additionally, the algorithm has an inherent tolerance against processor failures. We say that a processor fails if it stops at some time t_f to communicate with other processors.

Theorem 2.21 (Fault-Tolerance). *Suppose that processor l fails at time t_f, and that the communication network remains jointly strongly connected after the failure of processor l. Let $z^{[l]}(t_f)$ be the last query point computed by processor l and define $\gamma^{[l]}(t_f) = c^\top z^{[l]}(t_f)$. Then the query-points computed by all processors converge, i.e., $\lim_{t\to\infty} \|z^{[i]}(t) - \bar{z}_{-l}\| \to 0$, with $\bar{z}_{-l}$ satisfying*

$$\bar{z}_{-l} \in \left(\bigcap_{i\neq l} \mathcal{Z}_i\right) \quad \textit{and} \quad c^\top \bar{z}_{-l} \leq \gamma^{[l]}(t_f).$$

Proof. Consider the evolution of the algorithm starting at time t_f. With Lemma 2.18 and Lemma 2.19 one can conclude that for all processors $i \neq l$, the query points will converge to the set $\left(\bigcap_{i\neq l} \mathcal{Z}_i\right)$. Additionally, the out-neighbors of the failing processor l have received a basis $B^{[l]}(t_f)$ such that the optimal value of the linear approximate program (2.3) is $\gamma^{[l]}(t_f)$. Any query point $z^{[i]}(t), t \geq t_f$, subsequently computed by the out-neighbors of processor l as the solution to (2.3) must therefore be such that $c^\top z^{[i]}(t) \leq \gamma^{[l]}(t_f)$ for all $t \geq t_f$. □

This last result provides directly a paradigm for the design of fault-tolerant systems.

Corollary 2.22. *Suppose that for all $l \in \mathbf{V}$, $\bigcap_{i=1,i\neq l}^{n} \mathcal{Z}_i = \mathcal{Z}$. Then for all $l \in \mathbf{V}$, $\bar{z}_{-l} = z^*$ with z^* the optimal solution to* (2.1).

Up to now, we have assumed that the original problem (2.1) is neither infeasible nor unbounded. However, the cutting-plane consensus algorithm can be easily modified to handle also those exceptional cases, as summarized in the following proposition.

Proposition 2.23 (Exception Handling). *Consider a network of processors performing the cutting-plane consensus algorithm.*

1. *If $\mathcal{Z} = \emptyset$, there is a finite time instant at which the linear approximate problem* (2.3) *computed by some processor is infeasible.*

2. *If $\max_{z\in\mathcal{Z}} c^\top z \nless \infty$, the limit points $\bar{z}^{[i]}$ are on some "artificial" cutting-plane defined for the initial basis $B_0^{[i]}$.*

Proof. To prove statement (i), we have to show that $\bigcap_{i=1}^{n} \mathcal{Z}_i = \emptyset$ implies that there exists a finite time instant T and a processor $j \in V$ such that the linear approximate problem (2.3) computed by processor j at time T is infeasible. Assume, in order to get a contradiction, that $\bigcap_{i=1}^{n} \mathcal{Z}_i = \emptyset$, and for all $i \in \{1, \ldots, n\}$ and all $t \geq 0$ the linear approximate problem (2.3) is feasible. Then from Lemma 2.18 follows that each sequence $\{z^{[i]}(t)\}$ converges to a limit point $\bar{z}^{[i]} \in \mathcal{Z}_i$ and from Lemma 2.19 that all limit points are identical, i.e., $z^{[1]} = \ldots = \bar{z}^{[n]}$. This contradicts the assumption $\bigcup_{i=1}^{n} \mathcal{Z}_i = \emptyset$, and the statement follows.

The cutting-plane consensus algorithm is initialized with "artificial" cutting-planes $B_0^{[i]}$ such that $\max_{z\in\mathcal{B}^{[i]}(0)} c^\top z < \infty$. From Lemma 2.17 (iv) follows that all query points $z^{[i]}(t)$ generated by the cutting-plane consensus algorithm satisfy $c^\top z^{[i]}(t) \leq \max_{z\in\mathcal{B}^{[i]}(0)} c^\top z < \infty$. Now, if the problem is unbounded, the query points must diverge. However, they cannot exceed the artificial cutting-planes. Thus, the query points must always lie on some artificial cutting-plane contained in $\bigcup_{i=1}^{n} B^{[i]}(0)$. □

The abstract problem formulation (2.1) and the cutting-plane consensus algorithm provide a *general framework for distributed convex optimization.* We show in the following that a variety of important representations of the constraint sets are covered by this set-up. Depending on the formulation of the local constraint sets $\mathcal{Z}_i$ different cutting-plane oracles must be defined, leading to different realizations of the algorithm.

2.4 Convex Inequality Constraints

Up to now, we discussed the cutting-plane consensus algorithm in an abstract description and assumed only convexity of the constraint sets without specifying its formal representation. Now, we want to be more specific and formally present the algorithm for certain representations of the constraint sets. We start the discussion with the most intuitive set-up, where the constraint sets are simply defined by (possibly nonlinear and non-differentiable) inequalities. In such a set-up, our algorithm becomes a distributed version of classical cutting-plane methods. In addition, we show that for linear programs, i.e., having only linear inequality constraints, the cutting-plane consensus algorithm turns out to be a *distributed simplex algorithm.*

2.4.1 Problem Formulation

The first problem formulation we discuss here, is the most intuitive realization of the general problem formulation (2.1). We assume in the following that the local constraint sets are defined by convex inequalities, i.e.,

$$\mathcal{Z}_i = \{z : f_i(z) \leq 0\}. \tag{2.23}$$

The functions $f_i : \mathbb{R}^d \mapsto \mathbb{R}$ are assumed to be convex, but need not to be differentiable. Please note that the optimization problem, we discuss in this section takes the form

$$\begin{aligned} &\text{maximize } c^\top z \\ &\text{subj. to } f_i(z) \leq 0, \quad i \in \{1, \dots, n\}. \end{aligned} \tag{2.24}$$

The problem formulation (2.24) includes also the case that processor is assigned more that one constraint. Let, for example, the constraint set be defined by several inequalities, i.e., $\mathcal{Z}_i = \{z : f_{i1}(z) \leq 0, f_{i2}(z) \leq 0, \dots, f_{ik}(z) \leq 0\}$. As differentiability of the constraint is not required, one can combine all these constraints into one convex constraint of the form

$$f_i(z) := \max_{j \in \{1,\dots k\}} f_{ij}(z) \leq 0.$$

Now, the problem can be clearly formulated in the form (2.24).

We need to recall now the concept of a *subdifferential* for the constraints. Given a point $z_q \in \mathbb{R}^d$, the subdifferential of the function f_i at z_q is defined as

$$\partial f_i(z_q) = \{g_i \in \mathbb{R}^d : f_i(z) - f_i(z_q) \geq g_i^\top (z - z_q), \ \forall z \in \mathbb{R}^d\}.$$

The subdifferential $\partial f_i(z_q)$ is nonempty if f_i is continuous. An element $g_i \in \partial f_i(z_q)$ is called a *subgradient* of f_i at z_q. If the function f_i is differentiable, then its gradient $\nabla f_i(z_q)$ is a subgradient. If $f_i(z) := \max_{j \in \{1,\dots k\}} f_{ij}(z)$, then $\partial f_i(z_q) = \mathbf{Co} \cup \{\partial f_{ij}(z_q) : f_{ij}(z_q) = f_i(z_q)\}$, where **Co** denotes the convex hull. A cutting-plane oracle for constraints of the form (2.24) is now as follows.

Cutting-plane Oracle: Given a query point z_q. If

1. $f_i(z_q) \leq 0$ return empty h;
2. $f_i(z_q) > 0$, return

$$h := \{z : f_i(z_q) + g_i^\top (z - z_q) \leq 0, \} \tag{2.25}$$

for some $g_i \in \partial f_i(z)$.

This is the classical cutting-plane oracle as introduced in Kelley (1960). To see that (2.25) is a cutting plane oracle, note that by the definition of a subgradient $f_i(z_q) + g_i^\top (z - z_q) \leq 0$ for all z satisfying $f_i(z) \leq 0$, while $f_i(z_q) + g_i^\top (z_q - z_q) = f_i(z_q) > 0$. It remains to verify Assumption 2.1. Assumption 2.1 (i) is in fact an assumption on the functions $f_i(z)$. It requires that the subgradients are bounded, i.e., $\|g_i\| < \infty$. This assumption is met in particular if f_i is Lipschitz, i.e.,

$$|f_i(z) - f_i(z')| \leq L_i \|z - z'\|_2$$

for some finite L_i. Then, it holds that $\|g_i\|_2 \leq L_i$ for all $g_i \in \partial f_i(z_q)$ and all z_q. Assumption 2.1 (ii) is satisfied, since $s(z_q) = f_i(z_q) + g_i^\top (z_q - z_q) = f(z_q)$, and $f(z_q) = 0$ implies $z_q \in \mathcal{Z}_i$.

We can conclude that the cutting-plane consensus algorithm is directly applicable to any problem of the form (2.24), as long as subgradients to the constraint functions can be obtained. It is easy to modify Algorithm (2.3.3) for such problems, by simply defining (S3) according to the cutting plane oracle proposed here. We discuss next some special problem, that clearly deserve particular attention.

2.4.2 Semidefinite Constraints

An important class of constraints are *semidefinite* constraints. Let $F_{ij} \in \mathbb{R}^{p \times p}$, $j \in \{1, \ldots d\}$ be real symmetric matrices. A semidefinite constraint takes the form

$$F_i(z) := F_{i0} + z_1 F_{i1} + \cdots + z_d F_{id} \preceq 0,$$

where "$\preceq 0$" means *negative semidefinite.* The constraint set $\mathcal{Z}_i = \{z : F_i(z) := F_{i0} + z_1 F_{i1} + \cdots + z_d F_{id} \preceq 0\}$, is clearly a closed convex set. To show the applicability of the cutting-plane consensus algorithm to semidefinite constraints, we only have to establish the existence of a cutting-plane oracle. After our previous discussion, it is little surprising that a cutting plane oracle exists if one can compute subgradients to the semidefinite constraints. How to compute subgradients for semidefinite constraints is well-known, see e.g., Scherer and Weiland (2004), and we briefly review it here. In a first step, we note that the semidefinite constraint $F_i(z) \preceq 0$ can be equivalently expressed as a scalar inequality constraint

$$f_i(z) := \lambda_{\max}(F_i(z)) \leq 0,$$

where $\lambda_{\max}(F_i(z))$ denotes the largest eigenvalue of $F_i(z)$. Now, the latter constraint can be equivalently written as

$$f_i(z) = \sup_{v^\top v = 1} v^\top F(z) v \leq 0.$$

Given a query point z_q, one can determine v_q^* as the normalized eigenvector of $F_i(z_q)$ corresponding to $\lambda_{\max}(F_i(z_q))$, i.e., $v_q^* = \arg\max_{v^\top v=1} v^\top F(z_q)v$. Consequently, we can determine the value of the scalar constraint at a query point as $f_i(z_q) = v_q^{*\top} F(z_q)v_q^* = \lambda_{\max}(F_i(z_q))$. To obtain now a sub-gradient, we make the following considerations

$$\begin{aligned} f_i(z) - f_i(z_q) &= \max_{v^\top v=1} v^\top F_i(z)v - v_q^{*\top} F_i(z_q)v_q^* \\ &\geq v_q^{*\top}(F_i(z) - F_i(z_q))v_q^* = \sum_{j=1}^{d} \left(v_q^{*\top} F_{ij} v_q^*\right)(z_j - z_{q,j}), \end{aligned}$$

where the last equality follows from the affine structure of the semidefinite constraint.

We can conclude, that given a query point z_q and the normalized eigenvector v_q^* of $F_i(z_q)$ corresponding to $\lambda_{\max}(F_i(z_q))$, then the vector

$$g_i = [v_q^{*T} F_1 v_q^*, \ldots, v_q^{*T} F_d v_q^*]^\top$$

is a subgradient of $f_i(z)$ and equivalently to $F_i(z)$. This provides a constructive way to find subgradients to semidefinite constraints for the price of computing the eigenvector of a matrix. Since subgradients for semidefinite constraints can be easily computed, it follows directly that the cutting-plane consensus algorithm can also handle semidefinite constraints. The algorithm can therefore be seen in the context of the recent work on cutting-plane methods for semidefinite programming, such as Krishnan and Mitchell (2006), Konno et al. (2003). However, as discussed, our method is fully distributed and works in asynchronous communication networks.

2.4.3 Linear Constraints

Since the cutting-plane consensus algorithm relies heavily on the linear approximation of a convex optimization problem, it is natural to study the algorithm for the case that the original problem is already linear. For simplicity, we assume that each constraint set is defined by one linear inequality, i.e., $\mathcal{Z}_i = \{z : a_i^\top z \leq b_i\}$, and the global feasible set is the polyhedron

$$\mathcal{Z} = \{z : a_i^\top z \leq b_i, i = 1, \ldots, n\}. \tag{2.26}$$

It is not difficult to generalize the discussion to the situation, where the local constraint sets $\mathcal{Z}_i$ are general polyhedra.

The question we want to investigate now is how the cutting-plane consensus algorithm looks like for a linear program

$$\begin{aligned} &\text{maximize} \quad c^\top z \\ &\text{subj. to} \quad a_i^\top z \leq b_i, \quad i \in \{1, \ldots, n\}. \end{aligned} \tag{2.27}$$

We can, for example, use the cutting-plane oracle proposed in (2.25), for the constraints $f_i(z) = a_i^\top z - b_i \leq 0$. It can be readily seen that the subgradient to the linear constraint is simply $g_i = a_i$. The cutting-plane computes as

$$f_i(z_q) + g_i^\top (z - z_q) = a_i^\top z - b_i \leq 0,$$

i.e., it is exactly the original linear constraint. Now, the polyhedral outer approximation of the feasible set $\mathcal{Z}$ is simply defined by a subset of the half-spaces defining $\mathcal{Z}$. Let now the lexicographically minimal solution be considered as a unique solution rule in the cutting-plane consensus algorithm. Then, a basis of the approximate program corresponds always the solution that is at a *vertex* of the polyhedron $\mathcal{Z}$. Thus, the basis update in the cutting-plane consensus algorithm, applied to a linear program, corresponds to moving from one vertex of $\mathcal{Z}$ to a new vertex that has a better cost. This is exactly the idea of a *simplex algorithm* (Dantzig, 1963).[3] The simplex algorithm is widely considered to be one of the most important algorithm in mathematical programming, see e.g. Dongarra and Sullivan (2000), and is a standard tool for solving linear programs. Now, if cutting-plane consensus algorithm is applied to linear programs and the lexicographic optimal solution is considered, the algorithm can be interpreted as a distributed version of the simplex algorithm. It can be informally described as follows:

Distributed Simplex Algorithm: A processor stores a set of linear constraints, forming a basis for the linear program. Each processor repeats the following steps:

(S1) it receives the basis of its in-neighbors;

(S2) it computes the lexicographic optimal solution to the newly formed linear program;

(S3) if this solution violates its own original constraint, it includes its own constraint in the basis.

Simplex algorithms are known to be very efficient for solving linear programs in practice (although their worst case complexity is exponential), and they are in fact guaranteed to converge after a finite number of iterations. Now, a similar result can be established for the distributed simplex algorithm.

Proposition 2.24. *Consider a distributed linear program of the form* (2.27) *and let a processor network perform the Distributed Simplex algorithm. Then, there exists a finite time T, at which all processors have computed the same optimal solution to* (2.27).

Proof. The proof proceeds as the proof of Theorem 2.20, and exploits that a linear program admits only a finite number of bases and no basis can be repeatedly appear at one processor. □

We can conclude with the observation that the cutting-plane consensus algorithm (and the Distributed Simplex as its variant) solves linear programs very efficiently. The proposed Distributed Simplex is intimately related to the algorithm proposed in Notarstefano and Bullo (2011), for abstract optimization problems. The algorithm is presented here in the broader context of distributed convex optimization using cutting-planes. A dual version of this Distributed Simplex algorithm has been proposed in Bürger et al. (2012b). In this publication, the algorithm does not work with constraints but rather with "columns" of the dual linear program (i.e., a linear program of the form (2.5)). Additionally, the version of the Distributed Simplex Algorithm proposed in Bürger et al. (2012b) does not require to

[3] A simplex algorithm moves from from one feasible vertex of the constraint set to a "neighboring" vertex. In the context of the cutting-plane consensus algorithm, this corresponds to a situation, where at each iteration exactly one constraint is added.

solve an optimization problem in the local computations, but rather uses only algebraic pivot iterations to compute the lexicographic optimal solution. Additionally, it is discussed in Bürger et al. (2012b) that the use of lexicographic minimal solutions enables one to solve some classes of integer constrained problems. Since the lexicographic optimal solution is always located at some vertex of the constraint set, one can use it to compute integer solutions to the convex relaxations of certain problems. This property turns out to be of practical importance, for example, for solving the multi-agent assignment problem. For a detailed discussion of the Distributed Simplex Algorithm and its ability to solve multi-agent assignment problems, we refer to Bürger et al. (2012b).

2.4.4 Application Example: Position Estimation in Wireless Sensor Networks

As we have seen, the cutting-plane consensus algorithm is directly applicable to optimization problems where processors are assigned convex, possibly non-differentiable, inequality constraints. As an example application, we consider here the distributed position estimation problem in wireless sensor networks. In fact, we show that the problem can be readily formulated in the form (2.1), and that the cutting-plane consensus algorithm is particularly well suited to solve the problem.

Wide-area networks of cheap sensors with wireless communication are envisioned to be key elements of modern infrastructure systems. In most applications, only few sensors are equipped with localization tools, and it is necessary to estimate the position of the other sensors, see Bachrach and Taylor (2005). In Doherty et al. (2001) the sensor localization problem is formulated as a convex optimization problem, which is then solved by a central unit using semidefinite programming. The semidefinite formulation proposed in Doherty et al. (2001) has been later extended in the literature. A decomposition heuristic for the semidefinite program arising in position estimation problems has been proposed in Biswas and Ye (2006). The decomposition heuristic is motivated by the numerical problems of interior point algorithms with many semidefinite constraints and requires an a-priori clustering or the processors as well as a coordinating unit. Another distributed heuristic for position optimization has been proposed in Srirangarajan et al. (2008). The distributed approaches are not studied in the context of general distributed optimization. We formulate the distributed position estimation problem given in Doherty et al. (2001) in the general distributed convex optimization framework (2.1) and show that the general cutting-plane consensus algorithm can be used for a fully distributed solution, using only local message passing between the sensors.

Let in the following $\mathbf{v}_i \in \mathbb{R}^2$ denote the known position of sensor $i \in \{1, \dots, n\}$. We want to estimate the unknown position of an additional sensor $z \in \mathbb{R}^2$. In Doherty et al. (2001), two different estimation mechanisms are considered:

- laser transmitters at nodes which scan through some angle, leading to a cone set, which can be expressed by three linear constraints of the form $f(z) := a_i^\top z - b_i \leq 0$, $a_i \in \mathbb{R}^{2\times 1}$ and $b_i \in \mathbb{R}$, two bounding the angle and one bounding the distance;

- the range measurement of the RF transmitter, leading to circular constraints of the form $\|z - \mathbf{v}_i\|_2^2 \leq r_i^2$. Using the Schur-complement, the quadratic constraint can be

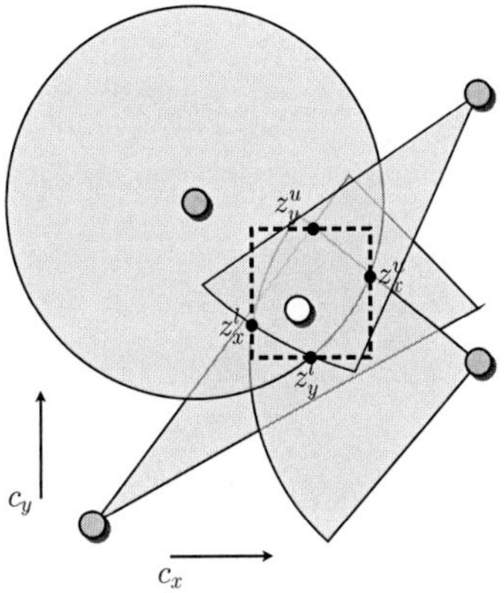

Figure 2.3: Localization of the white node by set estimates of the four gray nodes. The set estimate is given by the bounding box which is determined by the four points $z_x^l, z_x^u, z_y^l, z_y^u$. The four extreme points can be found with the cutting-plane consensus algorithm.

formulated as a semidefinite constraint of the form

$$F_i(z) := (-1)\begin{bmatrix} r_i I_2 & (z - \mathbf{v}_i) \\ (z - \mathbf{v}_i)^\top & r_i \end{bmatrix} \preceq 0,$$

where I_2 is the 2×2 identity matrix.

Each sensor i can bound the position of the unknown sensor to be contained in the convex set $\mathcal{Z}_i$, which is, depending on the available sensing mechanism, a disk represented by

- a semidefinite constraint $\mathcal{Z}_i = \{z : F_i(z) \preceq 0\}$,
- a cone $\mathcal{Z}_i = \{z : f_{ij}(z) \leq 0, j = 1, 2, 3\}$, or
- a quadrant, $\mathcal{Z}_i = \{z : F_{ij}(z) \preceq 0, f_{ij} \leq 0, j = 1, 2, 3\}$.

The considered problem for the sensing nodes is now to compute the smallest box

$$\{z \in \mathbb{R}^2 : [z_x^l, z_x^l]^\top \leq z \leq [z_x^u, z_y^u]^\top\}$$

that is guaranteed to contain the unknown position. As proposed in Doherty et al. (2001), the minimal bounding box can be computed by solving four optimization problems with linear objectives. To compute, for example, z_x^u one defines the objective $c_x = [1, 0]^\top$ and solves

$$z_x^u := \max\ c_x^\top z,\ \text{s.t. } z \in \bigcap_{i=1}^{n} \mathcal{Z}_i.$$

In the same way z_x^l, z_y^l, z_y^u can be determined. Figure 2.3 illustrates a configuration where four nodes estimate the position of one node.

2.5 Robust Optimization with Uncertain Constraints

We move now to a more general and less intuitive representation of the constraint sets. In fact, the general problem formulation (2.1) includes explicitly distributed robust optimization problems with uncertain constraints. The cutting-plane consensus algorithm turns out to be an efficient tool for solving *robust optimization problems in peer-to-peer processor networks.*

2.5.1 Problem Formulation

In particular, we consider distributed robust optimization problems of the form

$$\begin{aligned} &\text{maximize} \quad c^\top z \\ &\text{subj. to} \quad f_i(z,\theta_i) \leq 0, \quad \forall \theta_i \in \Omega_i,\ i \in \{1,\dots,n\}, \end{aligned} \tag{2.28}$$

where θ_i is an uncertain parameter, taking values in the compact convex set Ω_i. The constraint set is now given by

$$\mathcal{Z}_i = \{z : \ f_i(z,\theta) \leq 0, \text{ for all } \theta \in \Omega_i\}. \tag{2.29}$$

We assume that f_i is convex in z for any fixed θ. This assumption ensures that $\mathcal{Z}_i$ is a convex set. Sometimes, the problem (2.28) is called convex, if additionally f_i is concave in θ and Ω_i is a convex set (Lopez and Still, 2007). As we will see later on, the second condition will ensure that the problem can be solved exactly by our algorithm.

The problem (2.1) with constraints of the form (2.28) is a distributed *deterministic robust* (Ben-Tal et al., 2009) or distributed *semi-infinite* optimization problem (Lopez and Still, 2007). Each processor has knowledge of an infinite number of constraints, determined by the parameter θ and the uncertainty set Ω_i. Obviously, uncertain constraints as (2.28) appear frequently in distributed decision problems. Here we focus on a deterministic worst-case optimization problem, where a solution that is feasible for any possible representation of the uncertainty is sought. The centralized theory of robust optimization has a long history, see Dantzig (1955), and a comprehensive theory has been developed for centralized systems, see, e.g., Ben-Tal et al. (2009). A comprehensive theory for robust optimization in centralized systems has been developed and is presented, e.g., in Ben-Tal et al. (2009).

Nowadays, mainly two different approaches are pursued in robust optimization. A first research direction aims at formulating robust counterparts of the uncertain constraints (2.28), leading often to nominal semidefinite problems, see e.g., Ben-Tal et al. (2009). In another research direction the infinite, uncertain constraints are replaced by a finite number of "sampled" constraints. Sampling methods select a finite number of parameter values and provide bounds for the expected violation of the uncertain constraints, see e.g., Calafiore (2010). In a distributed setup, a sampling approach has been explored in Carlone et al. (2012). Handling the uncertain constraints from a semi-infinite optimization point of view (2.28), allows us also to apply exchange methods (Reemtsen, 1994), where the sampling point is chosen as the solution of a finite approximation of the optimization problem. Recently, cutting-plane methods have been considered in the context of centralized robust optimization in Mutapcic and Boyd (2009). Robust optimization in processor networks is a relatively new problem. In Yang et al. (2008), robust optimization for communication networks using dual decomposition is considered.

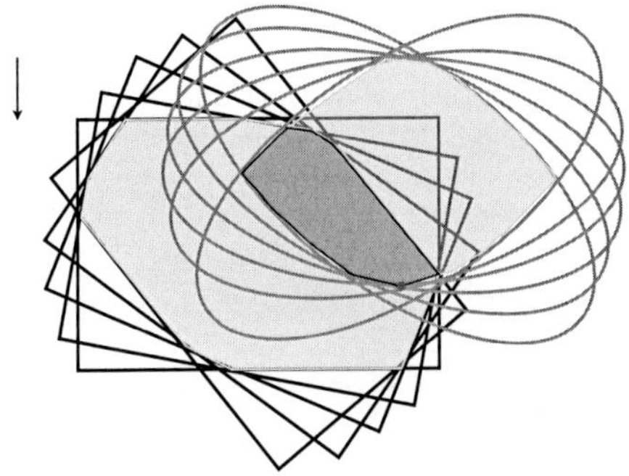

Figure 2.4: Illustration of an optimization problem with semi-infinite constraints. Here, the rotation of the constraint sets is an uncertainty. Although the basic constraints have a simple structure, the shape of the feasible set (dark) is non-trivial. The optimal solution in the intersection of the rotated constraints is indicated by the single marked point.

We connect the robust optimization problem with uncertain constraints (2.28) to our general distributed optimization framework (2.1), and show that the cutting-plane consensus algorithm can solve the problem in processor networks. The challenge for the application of the cutting-plane consensus algorithm is again to define a cutting plane oracle for the constraint sets (2.29). We propose a cutting-plane oracle for the distributed robust optimization problem (2.28) as follows.

Pessimizing Cutting-Plane Oracle: Given a query point z_q. Compute the worst-case parameter value θ_q^* as

$$\theta_q^* = \arg\max_{\theta} f_i(z_q, \theta) \quad \text{s.t. } \theta \in \Omega_i. \tag{2.30}$$

If

1. $f_i(z_q, \theta_q^*) \leq 0$ then $z_q \in \mathcal{Z}_i$ and return empty h;
2. $f_i(z_q, \theta_q^*) > 0$ then $z_q \notin \mathcal{Z}_i$ and return

$$h := \{z : f_i(z_q, \theta_q^*) + g_i^\top (z - z_q) \leq 0\}, \tag{2.31}$$

where $g_i \in \partial f_i(z_q, \theta_q^*)$ is a subgradient of f_i at z_q and θ_q^*.

To see that (2.31) is a cutting-plane, note that a query point $z_q \notin \mathcal{Z}_i$ is cut off, since $f_i(z_q, \theta_q^*) + g_i^\top (z_q - z_q) = f_i(z_q, \theta_q^*) > 0$. Additionally, for any point $z \in \mathcal{Z}_i$, we have $0 \geq f_i(z_i, \theta)$ for all $\theta \in \Omega_i$, and in particular $0 \geq f_i(z_i, \theta_q^*) \geq f_i(z_q, \theta_q^*) + g_i^\top (z - z_q)$. In order to meet Assumption 2.1, we assume again that the sub-gradients are bounded. Again, this is given if f_i satisfies some Lipschitz condition. Additionally, it can be directly seen that Assumption 2.1 (ii) is satisfied since $f_i(z_q, \theta_q^*) \to 0$ implies that $z_q \to \mathcal{Z}_i$. To facilitate the understanding of the cutting-plane consensus algorithm applied to robust optimization problems, we present in Algorithm 2 the formal description of the algorithm. Please note that the difference to the original cutting-plane consensus algorithm is purely in (S3), i.e., in the definition of the cutting-plane oracle.

Algorithm 2 Robust cutting-plane consensus

Processor state: $B^{[i]}(t)$

Initialization: $B^{[i]}(0) := B_M$

Function MSG: **Return** $B^{[i]}(t)$;

Function STF: **Set** $Y^{[i]}(t) := \cup_{j \in \mathcal{N}_I(i,t)} \texttt{MSG}(B^{[j]}(t))$;

(S1) $H_{tmp}^{[i]}(t) \leftarrow B^{[i]}(t) \cup Y^{[i]}(t)$;

(S2) $z^{[i]}(t) \leftarrow$ unique solution to $\min_{z \in \mathcal{H}_{tmp}^{[i]}(t)} c^\top z$

$B_{tmp}^{[i]}(t) \leftarrow$ set of active constraints at $z^{[i]}(t)$;

(S3) (i) $\theta_q^* \leftarrow \arg\max_\theta \; f_i(z^{[i]}(t), \theta) \quad \text{s.t. } \theta \in \Omega_i$

(ii) $h(z^{[i]}(t)) \leftarrow f_i(z^{[i]}(t), \theta_q^*) + g_i^\top (z - z^{[i]}(t)) \leq 0,$

$g_i \in \partial f_i(z_q, \theta_q^*)$;

(S4) $B^{[i]}(t+1) \leftarrow$ Basis of $B_{tmp}^{[i]}(t) \cup h(z^{[i]}(t))$.

2.5.2 Efficiently Solvable Problems

The oracle of the robust optimization problem requires to solve an additional optimization problem for determining the worst case parameter (2.30). Following Mutapcic and Boyd (2009), we call this the *pessimizing step*. For the practical applicability of our algorithm it is important to stress that the pessimizing steps are performed in parallel on different processors. In general, the pessimizing step can be performed by solving the optimization problem

$$\max_\theta \; f_i(z^{[i]}(t), \theta) \quad \text{s.t. } \theta \in \Omega_i$$

with numerical tools. However, the previous optimization problem demarcates the boundary the between "easy" and "hard" robust optimization problems. If f_i is concave in the uncertain parameter θ, then the problem can be efficiently and exactly solved. Otherwise, global optimization methods are required and one can only expect to obtain an approximation of the robust optimal solution.

However, it seems worth to outline some important cases for which the pessimizing step can be efficiently solved. We follow in the next discussion the reference Mutapcic and Boyd (2009).

- **Scenario Uncertainties:** An uncertainty set Ω_i is said to be scenario generated (see Ben-Tal et al. (2009)), if it is given as the convex hull of several parameters (i.e., "scenarios")

$$\Omega_i = \mathbf{Co}\{\theta^1, \ldots, \theta^M\}.$$

Let z_q be a query point. If f_i is convex in θ_i for all z, then the pessimizing oracle must select the maximum of the finite set

$$\{f_i(z_q, \theta^1), \ldots, f_i(z_q, \theta^M)\}.$$

Thus, the pessimizing step can be solved exactly by evaluating and comparing a finite number of functions. Note that this "brute force" method applies in particular if the dependence on the uncertainty is quadratic, or if f_i represents a semidefinite constraint with an affine dependence on the uncertainty.

- **Polytopic Uncertainty:** Consider a polytopic uncertainty set

$$\Omega_i = \{\theta : G\theta \leq \ell\}.$$

If f_i has an affine dependence on θ_i, i.e., $f_i(z, \theta_i) = \alpha_i(z)\theta_i + \beta_i(z)$, the pessimizing step can be solved as a linear program. If f_i is convex, the polytopic uncertainty has to be treated as a scenario uncertainty.

- **Ellipsoidal Uncertainties:** Consider an ellipsoidal uncertainty set

$$\Omega_i = \{\theta : \theta = \bar{\theta}_i + P_i u, \ \|u\|_2 \leq 1\}$$

for some nominal parameter value $\bar{\theta}_i$ and a positive definite matrix P_i. If $f_i(z, \theta_i)$ is an affine function in θ_i, i.e., $f_i(z, \theta_i) = \alpha_i(z)\theta_i + \beta_i(z)$, then the worst-case parameter value can be computed analytically as

$$\theta_i^* = \bar{\theta}_i + \frac{P_i P_i^\top \alpha_i(z)}{\|P_i \alpha_i(z)\|_2}. \tag{2.32}$$

We can conclude with the statement that there are, in fact, various important situations, where the pessimizing step can be solved efficiently and exactly. In other cases, the pessimizing step can be solved approximately. Although one can then not get the optimal robust solution, the cutting-plane consensus algorithm is still of practical use.

2.5.3 Computational Study: Robust Linear Programming

We evaluate in the following the time complexity of the algorithm in a computational study for distributed robust linear programming. We follow here Ben-Tal and Nemirovski (1999) and consider robust linear programs in the form (2.28) with linear uncertain constraints

$$a_i^\top z \leq b_i, \quad a_i \in \mathcal{A}_i, \quad i \in \{1, \ldots, n\}. \tag{2.33}$$

The data of the constraints is only known to be contained in a set, i.e., $a_i \in \mathcal{A}_i$. Although our algorithm can in principle handle any convex uncertainty set $\mathcal{A}_i$, we restrict us for this computational study to the important class of *ellipsoidal uncertainties* $\mathcal{A}_i = \{a_i : a_i = \bar{a}_i + P_i u_i, \|u_i\|_2 \leq 1\}$. The uncertainty ellipsoids are centered at the points $\bar{a}_i$ and their shapes are determined by the matrices $P_i \in \mathbb{R}^{d \times d}$. It is known in the literature that the centralized problem can be solved as a nonlinear *conic quadratic program* (Ben-Tal and Nemirovski, 1999)

$$\max c^\top z, \quad \text{s.t.} \quad \bar{a}_i^\top z + \|P_i z\|_2 \leq b_i, \quad i \in \{1, \ldots, n\}. \tag{2.34}$$

We will apply our algorithm directly to the uncertain problem model and use the nonlinear problem formulation (2.34) only as a reference for the computational study. For the particular problem (2.33) the pessimizing step can be performed analytically. Note that $\sup_{a_i \in \mathcal{A}_i} a_i^\top z_q = \bar{a}_i^\top z_q + \sup_{\|u\|_2 \leq 1}\{u^\top P_i^\top z_q\} = \bar{a}_i^\top z_q + \|P_i^\top z_q\|_2$. The worst-case parameter is therefore given by

$$a_i^* = \bar{a}_i + \frac{P_i P_i^\top z_q}{\|P_i z_q\|_2}. \tag{2.35}$$

A cutting-plane defined according to (2.31) takes simply the form $a_i^* z \leq b_i$, i.e., the linear constraint with the worst case parameter value.

For the computational study, we generate random linear programs in the following way. The nominal problem data $a_i \in \mathbb{R}^d$ and $c \in \mathbb{R}^d$ are independently drawn from a Gaussian distribution with mean 0 and standard deviation 10. The coefficients of the vector b are then computed as $b_i = \left(a_i^\top a_i\right)^{1/2}$. This random linear program model has been originally proposed in Dunham et al. (1977). The matrices P_i are generated as $P_i = M_i^\top M_i$ with the coefficients of $M_i \in \mathbb{R}^{d \times d}$ chosen randomly according to a normal distribution with mean 0 and standard deviation 1. All simulations are done with dimension $d = 10$. We consider the number of communication rounds required until the query points of all processors are close to the optimal solution z^*, i.e., we stop the algorithm centrally if for all $i \in \mathbf{V}$, $\|z^{[i]}(t) - z^*\|_2 \leq 0.1$. In Figure 2.5, the completion time for two different communication graphs is illustrated. We compare random Erdös-Rényi graphs, with edge probability $p = 1.2\frac{\log(n)}{n}$, and circulant graphs with 5 out-neighbors for each processor. It can be seen in Figure 2.5 that the number of communication rounds grows with the network size for the circulant graph, which have a growing diameter, but remains almost constant for the random Erdös-Rényi graphs, which have always a small diameter.[4] The simulations suggests, that the completion time depends primarily on the *diameter* of the communication graph.

We consider for a comparison the ADMM algorithm combined with a dual-decomposition, as described, e.g., in (Boyd et al., 2010, pp. 48), to solve the nominal conic quadratic problem representation (2.34) of the robust optimization problem.[5] In one iteration of the ADMM algorithm, all processors must update their local variables synchronously and then compute the average of all decision variables. Figure 2.5 (right axis) shows the number of iterations of the ADMM to compute the solution to the random linear programs with the same precision as the cutting-plane consensus algorithm. Note that the ADMM algorithm requires almost three times more iterations than the cutting-plane consensus algorithm requires communication rounds. The distributed implementation of Boyd et al. (2010)

[4]See Appendix C for a discussion.

[5]We use in the simulations a step-size $\rho = 200$, see (Boyd et al., 2010, Chapter 7) for the notation. Please note that the choice of the step-size of the ADMM method has to be done heuristically. We have selected the step-size as the best step-size we found experimentally for the smallest problem scenario $n = 20$. Although the convergence speed of the ADMM method might improve with another step-size, in our experience most heuristic choices led to a significant deterioration of the performance.

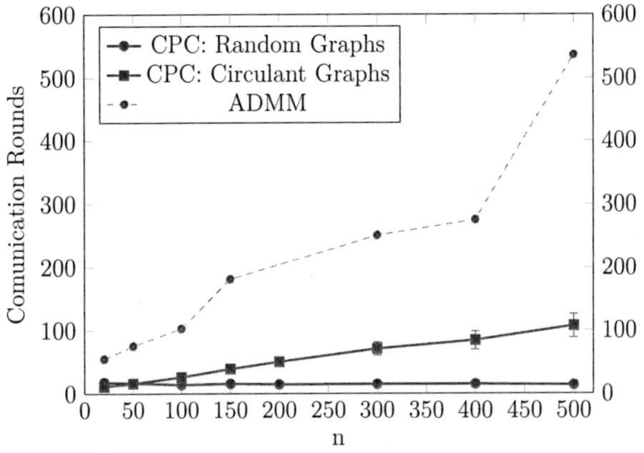

Figure 2.5: Average number of communication rounds and 95% confidence interval required to compute the optimal solution to randomly generated robust linear programs with a precision of $\epsilon < 0.1$ for Erdös-Rényi graphs (blue) and circulant graphs (red) with the cutting-plane consensus (CPC) algorithm. For comparison, the dashed line shows the number of iterations of the ADMM algorithm with dual decomposition (dashed line).

requires at each iteration an averaging of the local solutions. This can require a number of communication rounds in $\Omega\left(n^2 \log(\frac{1}{\delta})\right)$, where δ is the desired precision Olshevsky and Tsistiklis (2009). The consensus algorithm is not required by the distributed variants of Schizas et al. (2008), Wei and Ozdaglar (2012). However, these distributed implementations will require even more iterations of the algorithm.

Although the simulations do not compare the complexity of the algorithms in terms of computation units, they clearly suggest that the cutting-plane consensus algorithm is advantageous for applications where communication is costly or time consuming.

2.6 Conclusions

We presented in this chapter a novel distributed algorithm, the cutting-plane consensus algorithm, that can solve a variety of distributed optimization problems over asynchronous communication networks. We considered therefore a broad distributed optimization framework, where processor are assigned convex constraint sets and have to agree on the optimizer of a linear cost function. The proposed algorithm uses a polyhedral approximation method and unique solution linear programming as the main ingredients. The polyhedral approximation allows to reduce the complexity of the optimization problems that are solved between the processors. The use of unique solution linear programming techniques allows the processors to drop inactive constraints, such that the local memory requirements of the algorithm can be bounded, and ensures that the processors reach an agreement on the optimal solution.

Processors running the novel algorithm store and exchange a finite set of cutting-planes and compute, based on the current approximation of the constraint set, the next query point as a unique solution to the approximate linear program. We have formally proven that the processors reach asymptotically an agreement on an optimal solution of the global problem, and we have shown that the algorithm is extremely robust against failures of processors or of the communication network.

While the algorithm is originally presented in an abstract formulation for a general problem class, it was then specified to several important problem representations. For each problem representation, the algorithm takes a special structure. In this direction, we have shown how to solve convex optimization problems with convex inequality or semidefinite constraints. Additionally, we discussed that the novel algorithm becomes a distributed simplex algorithm, when it is applied to linear programs. Following this discussion, we have shown that the novel cutting-plane consensus algorithm can be easily adapted to solve robust optimization problems. For robust optimization, the algorithm has to determine repeatedly a worst-case parameter vector, which requires to solve an additional optimization problem. With this worst-case parameter, it is straight forward to design a suitable cutting-plane oracle for the uncertain constraints, and therefore to solve the robust optimization problem using the cutting-plane consensus algorithm.

In the next chapter, we will see that the cutting-plane consensus algorithm has a further important application area, namely distributed model predictive control problems. We will in the following consider the *dual* version of the algorithm and see that trajectory exchange method are variants of the cutting-plane consensus algorithm.

Chapter 3

Dual Cutting-Plane and Trajectory Exchange Optimization

3.1 Introduction

In this chapter, we show that the cutting-plane consensus algorithm has an intuitive and elegant interpretation, when it is applied to *distributed model predictive control* problems. It will, somehow surprising, turn out that the cutting-plane consensus algorithm then becomes a *trajectory exchange method*, where dynamical systems predict and exchange "plans" about their future behavior.

The discussion in this chapter is started with a basic control set-up, involving several physically independent dynamical systems that are coupled by a constraint on the systems output. Roughly speaking, the idea of model predictive control is to compute over a finite time horizon a set of (predicted) trajectories for all systems that satisfy the coupling constraints. The first piece of the computed input trajectory is then applied to the real physical system and based on the new initial condition, a new set of trajectories is predicted. This conceptual idea leads directly to a feedback control law. The first challenge in distributed model predictive control is to find at each time instant a set of feasible trajectories over a finite horizon, i.e., to *"[...] ensure that the distributed decision making leads to actions that are consistent with the actions of others and satisfy the coupling constraints"* (Richards and How, 2007). Thus, in the first place, the decision makers have to solve at each time instant a feasibility problem. However, once the feasibility problem is solved, the solution should be improved to satisfy, at best, some performance criteria. Thus the problem to be solved becomes a distributed optimization problem.

For solving the feasibility problem at each time instant, trajectory exchange methods have recently attained significant attention in the distributed model predictive control literature, see e.g., Grüne and Worthmann (2012); Müller et al. (2012); Pannek (2013); Richards and How (2007); Trodden and Richards (2010, 2013). The named trajectory exchange methods are, from an optimization perspective, heuristics that ensure feasibility of the local decisions but are not capable of finding an optimal solution. We will analyze and interpret these trajectory exchange methods here in the context of polyhedral approximations. This will provide us with new insights on why the methods fail to compute optimal solutions and how the algorithms might be redesigned.

In this chapter, we restrict the discussion to a convex framework and consider the predictive control problem for linear dynamical systems that are coupled by polyhedral constraints. In a first step, we derive an alternative formulation of the cooperative predictive control problem as an *almost separable convex optimization* problem in the output trajectories. Then, we use the theoretical framework established before, to explain why the basic trajectory exchange method, i.e., the method of Richards and How (2007), fails to compute an optimal solution to the optimal control problem. We show in particular that the basic trajectory exchange method can be interpreted as a variant of a basis exchange algorithm and therefore as a variation of the cutting-plane consensus algorithm. However, the method turns out to be under-parameterized (i.e., not enough trajectories are stored and exchanged by the systems), such that the method fails to compute the optimal solution.

Following this discussion, we focus on the application of the cutting-plane consensus algorithm to the class of almost separable optimization problems. In fact, the cutting-plane consensus algorithm can solve the *dual problem* of the almost separable optimization problem. In this context, the algorithm has several interpretations. For example, considering the primal formulation of the cutting-plane consensus algorithm applied to the almost separable problem, the algorithm turns out to be a variant of a *nonlinear, distributed Dantzig-Wolfe decomposition.* Additionally, if it is applied to a distributed predictive control problem, the algorithm turns out to be a trajectory exchange method, where processors exchange predictions about their behavior. However, in contrast to the original trajectory exchange methods, the novel algorithm ensures convergence to a central optimal solution, provided it is performed sufficiently long.

3.2 A Motivating Problem: Distributed Cooperative Model Predictive Control

We consider now a distributed model predictive control problem and investigate the problem in the light of the cutting-plane consensus algorithm presented in the previous chapter.

3.2.1 Problem Formulation

We consider in the following a group of n dynamical systems, with each system $i \in \{1, \dots, n\}$ being governed by a linear time-invariant discrete time dynamical system

$$x_i(t+1) = A_i x_i(t) + B_i u_i(t), \tag{3.1}$$

where $x_i(t) \in \mathbb{R}^{p_i}$ is the systems state and $u_i(t) \in \mathbb{R}^{q_i}$ is the control input. We assume that the states and the control inputs of the system are constrained to compact and convex sets containing the origin, i.e., $x_i(t) \in \mathcal{X}_i$, $u_i(t) \in \mathcal{U}_i$ for all $t \geq 0$. We associate to each system an output

$$y_i(t) = C_i x_i(t) + D_i u_i(t), \tag{3.2}$$

with $y_i(t) \in \mathbb{R}^r$ and consider a coupling among the systems by the constraint

$$\sum_{i=1}^{n} y_i(t) \leq h_t, \quad \text{for all } t \geq 0. \tag{3.3}$$

Note that this formulation includes all situations, where the sum of some output function is constrained to a polyhedral convex set. For a time interval $[t, t+T]$, we will write

$\mathbf{h} = [h_t^\top, \ldots h_{t+T-1}^\top]^\top$ and $\boldsymbol{y}_i = [y_i^\top(t), \ldots, y_i^\top(t+T-1)]^\top$. The constraint can now be rewritten in a more compact form as

$$\sum_{i=1}^{n} \boldsymbol{y}_i \leq \mathbf{h}.$$

In the following, we will adopt the notation to the predictive nature of the problem, and we will write $x_i(t+\tau|t)$ to denote the predicted state of system i as computed at time t for the future time-instant $t+\tau$. The same notation is used for the predicted inputs $u_i(t+\tau|t)$ and outputs $y_i(t+\tau|t)$. We assume that each system is assigned an objective function penalizing states and inputs at each time step. At each time t the cost functions are truncated at a finite time horizon $[t, t+T]$, leading to the *finite horizon* cost

$$F_i(x_i(t), \boldsymbol{u}_i) = \sum_{\tau=0}^{T-1} f_i(x_i(t+\tau|t), u_i(t+\tau|t)) + V_i(x_i(t+T|t)), \tag{3.4}$$

where $f_i(x_i(t+\tau|t), u_i(t+\tau|t))$ is called the stage-cost, and $V_i(x_i(t+T))$ the terminal cost. We assume here that both f_i and V_i are convex in their arguments. Note that there are various ways to formulate the finite-horizon optimal control problem appearing in a model predictive control scheme in such a way that recursive feasibility and stability can be guaranteed, and among the most popular ones is adding terminal constraints and/or terminal cost functions.[1] If we take now a *centralized* perspective, the finite horizon optimal control problem that has to be solved at time instant t is the coordination problem

$$\begin{aligned} \min \quad & \sum_{i=1}^{n} \Big(\sum_{\tau=0}^{T-1} f_i(x_i(t+\tau|t), u_i(t+\tau|t)) + V_i(x(t+T|t)) \Big) \\ \text{subj. to} \quad & \sum_{i=1}^{n} y_i(t+\tau|t) \leq h_{t+\tau}, \\ & x_i(t+\tau+1|t) = A_i x_i(t+\tau|t) + B_i u_i(t+\tau|t), \\ & y_i(t+\tau|t) = C_i x_i(t+\tau|t) + D_i u_i(t+\tau|t), \\ & x_i(t|t) = x_i(t),\ x_i(t+\tau|t) \in \mathcal{X}_i,\ u_i(t+\tau|t) \in \mathcal{U}_i \\ & \quad i \in \{1, \ldots, n\},\ \tau \in \{0, \ldots, T-1\}, \end{aligned} \tag{3.5}$$

where $x_i(t)$ is the "real" state of system i at time t.

Remark 3.1. *We ignore here the classical MPC problem of prolonging known trajectories after propagation of time. This problem is, however, in the scope of most literature on distributed model predictive control. We refer again to the relevant literature (i.e., Grüne and Worthmann (2012); Müller et al. (2012); Richards and How (2007)) and assume simply that some method is known. Our contribution here is a modification of the existing schemes for solving the finite horizon optimal control problems exactly in a distributed way.*

[1] We want to emphasize that the exact formulation of finite horizon optimal control problem is not in the scope of this work. We refer the interested reader to the extensive literature, e.g., Grüne and Worthmann (2012); Müller et al. (2012); Richards and How (2007) for a detailed treatment of this problem. For the purpose of the present work, we simply assume that a suitable formulation of the finite-horizon optimal control problem is known. The main contribution here is the algorithmic method to solve the finite horizon problem.

3.2.2 Dual Semi-Infinite Problem Representation

We discuss now an alternative representation of the cooperative finite horizon optimal control problem.

Almost Separable Form The first observation we make here is that the model predictive control problem (3.5) can be formulated as an almost separable optimization problem. It is well-known, that over the finite horizon the input trajectory of a linear (time-invariant) system determines the state and output trajectory in an affine manner, i.e.,

$$\boldsymbol{x}_i = \Gamma_i \boldsymbol{u}_i + \beta_i x_i(t),$$

and

$$\boldsymbol{y}_i = (C_i\Gamma_i + D_i)\boldsymbol{u}_i + C_i\beta_i x_i(t),$$

where $x_i(t)$ is the state of system i at initialization time t, and Γ_i, β_i are suitably defined. We can now define the set of feasible output trajectories for each system as

$$\mathcal{Y}_i := \left\{ \boldsymbol{y}_i \in \mathbb{R}^{r \cdot T} \mid \exists \boldsymbol{u} \in \mathcal{U}_i, \text{s.t.} \boldsymbol{y}_i = (C_i\Gamma_i + D_i)\boldsymbol{u} + C_i\beta_i x_i(t), \text{ and } \Gamma_i \boldsymbol{u} + \beta_i x_i(t) \in \mathcal{X}_i \right\}.$$

Clearly, all sets $\mathcal{Y}_i$, $i \in \{1, \dots, n\}$, are convex. In what follows, we will consider the feasible output trajectories $\boldsymbol{y}_i$ as decision variables. We can associate to any feasible output trajectory $\boldsymbol{y}_i \in \mathcal{Y}_i$, at least one input trajectory $\boldsymbol{u}_i \in \mathcal{U}_i$ generating this output, and without loss of generality we associate to each output trajectory $\boldsymbol{y}_i \in \mathcal{Y}_i$ the input trajectory with minimal cost, i.e.,

$$\begin{aligned} F_i(\boldsymbol{y}_i) = \min_{\boldsymbol{u}_i \in \mathcal{U}_i} & \; F_i(x_{i0}, \boldsymbol{u}_i) \\ & \boldsymbol{y}_i = (C_i\Gamma_i + D_i)\boldsymbol{u} + C_i\beta_i x_{i0}, \\ & \Gamma_i \boldsymbol{u} + \beta_i x_i(t) \in \mathcal{X}_i. \end{aligned} \tag{3.6}$$

Now, note that the restriction to linear time-invariant systems implies that any convex combination of feasible output trajectories, i.e., $\hat{\boldsymbol{y}}_i^t = t\boldsymbol{y}_i' + (1-t)\boldsymbol{y}_i''$, $t \in [0,1]$, is generated by the convex combination of the inputs $\hat{u}_i^t = t\boldsymbol{u}_i' + (1-t)\boldsymbol{u}_i''$. Thus, convexity of $F_i(x_{i0}, \boldsymbol{u}_i)$ implies directly convexity of $F_i(\boldsymbol{y}_i)$. Now, the model predictive control problem (3.5) can be expressed as an almost separable optimization problem in the output trajectories

$$\begin{aligned} \min_{\boldsymbol{y}_i} \quad & \sum_{i=1}^{n} F_i(\boldsymbol{y}_i) \\ \text{subj. to} \quad & \sum_{i=1}^{n} \boldsymbol{y}_i \leq \mathrm{h} \\ & \boldsymbol{y}_i \in \mathcal{Y}_i, \; i \in \{1, \dots, n\}. \end{aligned} \tag{3.7}$$

Semi-Infinite Dual Representation We show next, that the dual problem to the almost separable problem (3.7) can be expressed as a problem in the general form (2.1). We consider the *partial Lagrangian function* of (3.7) with $\pi \in \mathbb{R}^{r \cdot T}$ as the dual variable of the coupling constraint, i.e.,

$$\begin{aligned} \max_{\pi \geq 0} \; & \left\{ \min_{\boldsymbol{y}_i} \sum_{i=1}^{n} F_i(\boldsymbol{y}_i) + \pi^\top \Big(\sum_{i=1}^{n} \boldsymbol{y}_i - \mathrm{h} \Big) \right\} \\ & \boldsymbol{y}_i \in \mathcal{Y}_i. \end{aligned} \tag{3.8}$$

The dual problem can now be rewritten as

$$\max_{\pi \geq 0} -\mathbf{h}^\top \pi + \sum_{i=1}^{n} \left\{ \min_{\boldsymbol{y}_i \in \mathcal{Y}_i} F_i(\boldsymbol{y}_i) + \pi^T \boldsymbol{y}_i \right\}.$$

We introduce now simply a new variable ν_i, that has to satisfy

$$\nu_i = \min_{\boldsymbol{y}_i \in \mathcal{Y}_i} F_i(\boldsymbol{y}_i) + \pi^\top \boldsymbol{y}_i.$$

With this new decision variable, we can represent the dual problem to (3.7) as the semi-infinite optimization problem

$$\begin{aligned} \max_{\pi \geq 0, \nu_i} \quad & -\mathbf{h}^\top \pi + \sum_{i=1}^{n} \nu_i \\ & \nu_i \leq F_i(\boldsymbol{y}_i) + \pi^\top \boldsymbol{y}_i, \ \forall \boldsymbol{y}_i \in \mathcal{Y}_i. \end{aligned} \tag{3.9}$$

Introducing now the variable $z = [\pi^\top, \nu_1, \ldots, \nu_n]^\top \in \mathbb{R}^{r \cdot T + n}$ and the cost vector $c = [-\mathbf{h}^\top, \mathbf{1}_n^\top]^\top$, the problem (3.9) is, except for the positivity constraint on π, given in the standard form (2.1), with constraint sets

$$\mathcal{Z}_i := \{(\pi, \nu_i) : \nu_i \leq F_i(\boldsymbol{y}_i) + \pi^\top \boldsymbol{y}_i, \ \forall \boldsymbol{y}_i \in \mathcal{Y}_i\}.$$

Note that in this representation, the linear semi-infinite constraints are parameterized by all feasible output trajectories $\boldsymbol{y}_i \in \mathcal{Y}_i$ and the corresponding value of the optimal local control problem $F_i(\boldsymbol{y}_i)$ as defined by (3.6). We will exploit this representation of the problem in the following to understand why trajectory exchange methods cannot compute the optimal solution, while our cutting-plane consensus algorithm can do so.

3.3 Revisiting the Richards and How Algorithm

We turn our attention now to a well-known trajectory exchange method for distributed model predictive control. The most fundamental trajectory exchange method was proposed by Richards and How (2007). This method has been the basis for various improved trajectory exchange methods, such as for example Grüne and Worthmann (2012); Müller et al. (2012); Pannek (2013); Trodden and Richards (2010, 2013). However, neither the original method by Richards and How (2007), nor the subsequently proposed methods have the same performance as a centralized model predictive control scheme. The reason for this is that none of the methods is capable to compute the optimal solution to the finite-horizon optimal control problems (3.5). We review here, exemplary for all trajectory exchange methods, the algorithm of Richards and How (2007) and use our previous theoretical considerations to explain why those methods cannot recover the performance of a centralized model predictive control scheme. For clarity of presentation, we restrict our analysis here to the problem formulation (3.5), although in principle more general constraints can be handled. Consistently with the notation in the previous section, we denote output trajectories predicted over the finite time horizon $[t, t+T]$ with $\boldsymbol{y}_i = [y_i^\top(t), \ldots, y_i^\top(t+T-1)]^\top$. Additionally, we denote a *collection of output trajectories* with B.

For applying the algorithm of Richards and How (2007), the systems have to be ordered and they have to communicate sequentially according to their labels $i \in \{1, \dots, n\}$. That is, the agents are assumed to communicate according to a cycle graph. The basic idea of the algorithm is that the systems store a set of n output trajectories, each one representing the current predicted plan of one system. In a sequential order, starting with the first system, i.e., $i = 1$, one system solves locally an optimal control problem, assuming the other trajectories as given and fixed. Then, the currently active system replaces its own trajectory with the newly computed one and passes the new set of n output trajectories to the next system, i.e., to system $i + 1$. Once all systems have updated their trajectories, all systems implement the first piece of the control input sequence generating their current predicted output trajectory and restart the iterative procedure.

The local subproblems appearing at each time instant are as follows. Let a set of n feasible output trajectories be given, i.e., $B = (\bar{\boldsymbol{y}}_1, \dots, \bar{\boldsymbol{y}}_n)$. The subproblem solved by each processor is

$$
\begin{aligned}
\min \quad & \Big(\sum_{\tau=0}^{T-1} f_i(x_i(t+\tau|t), u_i(t+\tau|t))\Big) + V_i(x(t+T|t)) \\
\text{subj. to} \quad & \boldsymbol{y}_i + \sum_{j \neq i} \bar{\boldsymbol{y}}_j \leq \mathbf{h} \\
& x_i(t+\tau+1|t) = A_i x_i(t+\tau|t) + B_i u_i(t+\tau|t), \\
& y_i(t+\tau|t) = C_i x_i(t+\tau|t) + D_i u_i(t+\tau|t), \\
& x_i(t|t) = x_i(t),\ x_i(t+\tau|t) \in \mathcal{X}_i,\ u_i(t+\tau|t) \in \mathcal{U}_i \\
& \tau \in \{0, \dots, T-1\}.
\end{aligned} \tag{3.10}
$$

Each subproblem (3.10) has as decision variables only the input variables of the currently active system and as constraints the local operational constraint plus the "reduced" coupling constraint as an output constraint. By construction, the new set of output trajectories is again feasible for the coupling constraint. We can now summarize the distributed MPC algorithm of Richards and How (2007) as follows:

Trajectory-Exchange DMPC

1. Set $t = 0$ and find a set of n feasible trajectories $B = \{\boldsymbol{y}_1^0, \dots, \boldsymbol{y}_n^0\}$.
2. Apply the first element of the control inputs sequence corresponding to the output trajectories in B for each subsystem and increment t.
3. Find a feasible extension of the trajectories in $B = \{\bar{\boldsymbol{y}}_1, \dots, \bar{\boldsymbol{y}}_n\}$ for the shifted time horizon $[t, t+T]$.
4. For $i = 1, \dots, n$; let agent i
   (i) receive the set of feasible output trajectories B from system $i - 1$;
   (ii) solve sub-problem (3.10) and compute a new trajectory $\boldsymbol{y}_i$;
   (iii) update B by replacing the $\bar{\boldsymbol{y}}_i$ with the newly computed $\boldsymbol{y}_i$.
5. Go to Step (2).

Step 3 is done in Richards and How (2007) by applying a stabilizing control law in the terminal region. However, we will not further discuss this aspect here, but we will focus on the local computations performed at each time instant. It is well-known that the algorithmic method ensures recursive feasibility of all optimization problems and satisfaction of all constraints. However, it is also known that this method is in general not able to reproduce the performance of a centralized algorithm. Even if we would go from the sequential to an iterative implementation of the algorithm, that is if Step 4 is performed repeatedly within one time step and agent n communicates the set of feasible trajectories to agent 1 such that the procedure can be repeated, the centralized optimal solution will not be computed. We say that the algorithm is not *consistent.* A simple, almost trivial example, can illustrate this problem.

Example 3.2. *Consider a simple problem with two decision makers, each supervising only one decision variable* $\boldsymbol{y}_1 \in \mathbb{R}$ *and* $\boldsymbol{y}_2 \in \mathbb{R}$*, respectively. The decision makers are assigned the objective functions* $F_i = \frac{1}{2}(\boldsymbol{y}_i - 1)^2$*,* $i \in \{1,2\}$ *and are coupled by the constraint* $\boldsymbol{y}_1 + \boldsymbol{y}_2 \leq 1$*. Suppose now that the decision makers are initialized with the feasible but suboptimal solution* $\boldsymbol{y}_1 = 1$ *and* $\boldsymbol{y}_2 = 0$*. Now, if system 1 solves* (3.11)*, assuming that* $\boldsymbol{y}_2 = 0$ *is given, it will compute again* $\boldsymbol{y}_1 = 1$*. The same holds for system 2. Thus, the process is "blocked" and the process will not converge to the central optimal solution.*

This "blocking" is caused by several reasons, which we will discuss next. To gain a better understanding of the blocking phenomenon, we will next provide an interpretation of the trajectory exchange method as a *variation of the cutting-plane consensus algorithm.* At first we note that, similar to (3.7), we can formulate the local subproblem (3.10) as a convex optimization problem in the output trajectories, i.e.,

$$\begin{aligned} \min_{\boldsymbol{y}_i} \quad & F_i(\boldsymbol{y}_i) \\ \text{subj. to} \quad & \boldsymbol{y}_i + \sum_{j \neq i} \bar{\boldsymbol{y}}_j \leq \mathbf{h} \\ & \boldsymbol{y}_i \in \mathcal{Y}_i. \end{aligned} \tag{3.11}$$

Without altering the optimization problem, we can add the constant term $\sum_{j \neq i} F_j(\bar{\boldsymbol{y}}_j)$ to the cost function of (3.11). Now, the dual problem, derived as before from the partial Lagrangian function, takes the form

$$\begin{aligned} \max_{\pi \geq 0, \nu_i} \quad & -\mathbf{h}^\top \pi + \sum_{i=1}^{N} \nu_i \\ \text{s.t.} \ & \nu_i - \boldsymbol{y}_i^\top \pi \leq F_i(\boldsymbol{y}_i), \quad \forall \boldsymbol{y}_i \in \mathcal{Y}_i \\ & \nu_j - \bar{\boldsymbol{y}}_j^\top \pi \leq F_j(\bar{\boldsymbol{y}}_j), \quad \forall j \neq i. \end{aligned} \tag{3.12}$$

The algorithm of Richards and How (2007) can therefore be interpreted as follows: The processors store and exchange an outer approximation of the semi-infinite constraint set, i.e., the trajectories $B = (\bar{\boldsymbol{y}}_1, \ldots, \bar{\boldsymbol{y}}_n)$. Whenever a processor receives a set of trajectories, it computes a new constraint, i.e., a trajectory $\boldsymbol{y}_i$, and updates its set of constraints, i.e., the collection of output trajectories B.

Thus, solving the local subproblem (3.11) of the trajectory exchange method proposed by Richards and How (2007), is equivalent to solving a polyhedral approximation of the dual problem. However, solving the local subproblems in the form (3.11) corresponds to keeping the constraint corresponding to system i in its original semi-infinite form, and approximating only the constraints corresponding to the other systems. The next observation we want to make is that for the given optimal control problem, a set of feasible trajectories containing exactly one trajectory per system, suffices to form a basis. We will call in the following the tuple consisting of one output trajectory and the corresponding objective value $(\boldsymbol{y}_i^l, F_i(\boldsymbol{y}_i^l))$ a *plan* of system i.

Proposition 3.3. *Assume a set of n plans $B = \{(\boldsymbol{y}_1, F_1(\boldsymbol{y}_1)), \ldots, (\boldsymbol{y}_n, F_n(\boldsymbol{y}_n))\}$ is known, containing exactly one plan per system and satisfying the coupling constraint. Then, B forms a basis for the dual problem* (3.9).

Proof. The trajectories $\{\boldsymbol{y}_1, \ldots, \boldsymbol{y}_n\}$ are a feasible solution to the primal problem (3.7). The value of the primal problem (3.7) at this solution is $\sum_{i=1}^n F_i(\boldsymbol{y}_i)$. Consider now the finite approximation of the dual problem (3.9), defined by B. We have to show that the approximation of the dual problem has a finite feasible solution. From strong duality follows that the optimal dual solution must be such that the dual problem has the same value as the primal problem, i.e., $\sum_{i=1}^n \nu_i - \mathbf{h}^\top \pi = \sum_{i=1}^n F_i(\boldsymbol{y}_i)$. Furthermore, the dual solution must be feasible, i.e., $\nu_i - \boldsymbol{y}_i^\top \pi \leq F_i(\boldsymbol{y}_i)$, $i \in \{1, \ldots, n\}$ and $\pi \geq 0$. Clearly, a finite optimal solution to the approximation of the dual problem is $\pi = 0$ and $\nu_i = F_i(\boldsymbol{y}_i)$. Thus, the n plans define an approximation of the dual problem with a finite feasible solution. Furthermore, if any of the constraints defined by the plans in B is removed, the approximate problem becomes unbounded. Therefore the set of plans B forms a basis. □

Now, we can also observe that the trajectory exchange method is in fact a basis exchange method, and can therefore be seen as a variant of the cutting-plane consensus algorithm. However, to understand blocking problem associated to the trajectory exchange method, one has to note that not all bases of (3.9) can be characterized with n plans.

Proposition 3.4. *The combinatorial dimension of* (3.9) *is $d = n + r \cdot T$.*

This can be trivially seen as the dual problem (3.9) has d decision variables, i.e., $\pi \in \mathbb{R}^{r \cdot T}$ and $\nu_1, \ldots, \nu_n$. Thus, the proposed trajectory exchange method, which stores and updates one trajectory per system, represents an *under-parameterized* version of a polyhedral approximation method. As the combinatorial dimension is $d = n + r \cdot T$, we might expect that it is necessary to store and exchange up to $n + r \cdot T$ trajectories in order to converge to the central optimal solution.

We will further investigate this issue in the remainder of this chapter and will present and discuss an improved version of the trajectory exchange method. To develop this novel trajectory exchange method, we will first discuss the general application of the cutting-plane consensus algorithm to convex optimization problems in the general almost separable form (3.7). We will see that the cutting-plane consensus algorithm applied to these problems becomes a distributed, asynchronous version of the nonlinear *Dantzig Wolfe decomposition.*

3.4 Distributed Nonlinear Dantzig-Wolfe Decomposition

We forget for now about the predictive control interpretation of the previous sections and focus on the general class of *almost separable convex optimization problems* of the form

$$\begin{aligned} \min_{\boldsymbol{y}_i} \quad & \sum_{i=1}^{n} F_i(\boldsymbol{y}_i) \\ \text{s.t.} \quad & \sum_{i=1}^{n} G_i \boldsymbol{y}_i \leq \mathbf{h} \\ & \boldsymbol{y}_i \in \mathcal{Y}_i, \; i \in \{1, \ldots, n\}, \end{aligned} \tag{3.13}$$

where $\boldsymbol{y}_i \in \mathbb{R}^{m_i}$ is a decision vector governed by processor i, $F_i : \mathbb{R}^{m_i} \mapsto \mathbb{R}$ is a convex objective function processor i aims to minimize, and $\mathcal{Y}_i \subset \mathbb{R}^{m_i}$ is a convex set, defining the feasible region for the decision vector $\boldsymbol{y}_i$. We assume here that all sets $\mathcal{Y}_i$ are bounded, and that a Slater point exists. The local decision variables $\boldsymbol{y}_i$ are all coupled by a linear separable constraint, with matrices $G \in \mathbb{R}^{\rho \times m}$ and a right-hand side vector $\mathbf{h} \in \mathbb{R}^{\rho}$. The coupling linear constraint is of dimension ρ, and we assume here that ρ is small compared to the number of decision variables, i.e., $\rho \ll \sum_{i=1}^{n} m_i$.

We have already seen in Section 3.2.2 that the dual problem to (3.13) can be written in the semi-infinite form

$$\begin{aligned} \max_{\pi \geq 0, \nu_i} \quad & -\mathbf{h}^{\top} \pi + \sum_{i=1}^{n} \nu_i \\ & \nu_i \leq F_i(\boldsymbol{y}_i) + \pi^{\top} \boldsymbol{y}_i, \; \forall \boldsymbol{y}_i \in \mathcal{Y}_i, \quad i \in \{1, \ldots, n\}. \end{aligned} \tag{3.14}$$

Except for the non-negativity constraint on the multipliers, this problem representation is exactly in the form (2.1). A linear cost should be maximized over the intersection of convex sets, where each convex set is assigned to one processor. We can therefore apply the cutting-plane consensus algorithm to solve this problem class. Note that the non-negativity constraint will appear in the linear approximate program (2.3), but we can still compute the minimal 2-norm solution of the linear program, and the convergence proof for the cutting-plane consensus algorithm remains valid.

3.4.1 Distributed Constraint Generation

The main ingredient for applying the cutting-plane consensus algorithm is the definition of a *cutting-plane oracle*. Let a query point $z_q = [\pi_q^{\top}, \nu_{q,1}, \ldots, \nu_{q,n}]^{\top}$ be given. The query point z_q is contained in the set $\mathcal{Z}_i$ if and only if

$$\nu_{q,i} \leq F_i(\boldsymbol{y}_i) + \pi_q^{\top} G_i \boldsymbol{y}_i, \quad \forall \boldsymbol{y}_i \in \mathcal{Y}_i.$$

Of course this is equivalent to requiring that the optimal value of the local *subproblem*

$$\begin{aligned} \min \; & F_i(\boldsymbol{y}_i) + \pi_q^{\top} G_i \boldsymbol{y}_i \\ & \boldsymbol{y}_i \in \mathcal{Y}_i \end{aligned} \tag{3.15}$$

is not smaller than the component of the query point $\nu_{q,i}$. Please note that the subproblems (3.15) are constructed only from the problem data corresponding to the respective processor, plus the components of the query point π_q and $\nu_{i,q}$. Based on this method to test the feasibility of a query point, we can directly derive a cutting-plane oracle.

Constraint Generating Oracle: Given a query point $z_q = [\pi_q^\top, \nu_{1,q}, \ldots, \nu_{n,q}]^\top$. Consider the local subproblem

$$\min_{\boldsymbol{y}_i} F_i(\boldsymbol{y}_i) + \pi_q^\top G_i \boldsymbol{y}_i, \quad \text{s.t. } \boldsymbol{y}_i \in \mathcal{Y}_i. \tag{3.16}$$

Let $\bar{\boldsymbol{y}}_i$ denote the optimal solution vector and γ_i^* the optimal value of (3.16). If

1. $\nu_{i,q} \leq \gamma_i^*$ then $z_q \in \mathcal{Z}_i$ and return empty h;
2. $\nu_{q,i} > \gamma_i^*$ then $z_q \notin \mathcal{Z}_i$ and return

$$h := \{(\pi, \nu) \,:\, \nu_i - F_i(\bar{\boldsymbol{y}}_i) - \pi^\top G_i \bar{\boldsymbol{y}}_i \leq 0\}.$$

It can be easily seen that the proposed procedure is a correct cutting-plane oracle. Clearly, for any query point $(\pi_q, \nu_q) \notin \mathcal{Z}_i$ we have $\nu_{q,i} - F_i(\bar{\boldsymbol{y}}_i) - \pi_q^\top G_i \bar{\boldsymbol{y}}_i > 0$. Thus, the query point is made infeasible by the oracle. Additionally, $\nu_i - F_i(\bar{\boldsymbol{y}}_i) - \pi^\top A_i \bar{\boldsymbol{y}}_i \leq 0$ for all $(\pi, \nu) \in \mathcal{Z}_i$. It can also be directly seen that Assumption 2.1 holds. Since the set $\mathcal{Y}_i$ is bounded, we will always have $\|G_i \bar{\boldsymbol{y}}_i\| \leq \infty$ and consequently Assumption 2.1 (i) holds. Additionally, note that $s(z_q) = \nu_{q,i} - F_i(\bar{\boldsymbol{y}}_i) - \pi_q^\top G_i \bar{\boldsymbol{y}}_i$. Clearly, as $z_q \to \bar{z}$ and $s(z_q) \to 0$ it must hold that $\bar{z} \in \mathcal{Z}_i$.

The constraints generated by this cutting-plane oracle have a very clear interpretation. The linear part is simply the contribution of processor i to the coupling constraint plus a vector of length n having only one nonzero element, i.e., with entry one, at the position i. The constant term is simply the value of the local objective function, evaluated at the candidate solution $\bar{\boldsymbol{y}}_i$. We can apply now the cutting-plane consensus algorithm to the dual (3.14). Since all assumptions of Theorem 2.20 are satisfied, we can conclude from this result that the algorithm will asymptotically compute the optimal dual solution, i.e.,

$$\lim_{t \to \infty} \|\pi^{[i]}(t) - \pi^*\|_2 \to 0.$$

However, we are in the first place interested in the primal optimal solution $\boldsymbol{y}_i^*$ to (3.13). Therefore, we need to think about how to find the optimal primal solution, given convergence of the dual solution. Let in the following $\boldsymbol{y}^{[i]}(\ell)$ denote the optimal solution to subproblem (3.16), computed with the query-point $z^{[i]}(\ell)$, and let $\boldsymbol{y}^* = [\boldsymbol{y}_1^*, \ldots, \boldsymbol{y}_n^*]$ be the optimal primal solution to (3.13).

Proposition 3.5. *Assume the functions $F_i(\boldsymbol{y}_i)$ in* (3.13) *are essentially smooth and strictly convex for all $i \in \{1, \ldots, n\}$. Then at each processor $i \in \{1, \ldots, n\}$ the solutions of the local subproblems* (3.16) *converge to the optimal primal solution of the corresponding processor, i.e.,*

$$\lim_{\ell \to \infty} \|\bar{\boldsymbol{y}}^{[i]}(\ell) - \boldsymbol{y}_i^*\| \to 0, \text{ for all } i \in V.$$

Proof. As a consequence of Theorem A.6, the assumption that $F_i(\boldsymbol{y}_i)$ is essentially smooth and strictly convex implies that for any $\pi^{[i]}(\ell)$ the solution to the subproblem (3.16) is unique and is feasible with respect to the local constraint set $\mathcal{Y}_i$. Since all processors converge to the same optimal dual solution, i.e., $\lim_{\ell\to\infty}\pi^{[i]}(\ell)=\pi^*$ for all $i\in\{1,\dots,n\}$ (Theorem 2.20), the sequences of primal solutions $\bar{\boldsymbol{y}}^{[i]}(\ell)$ converge for all $i\in\{1,\dots,n\}$ such that the limit points combined satisfying the coupling constraint. Thus, limit points are primal feasible for (3.13) and from strong duality follows that they must be the primal optimal solution. □

However, in many important applications strict convexity of $F_i(\cdot)$ is not given. In fact, recovering a primal optimal solution from the dual solution is more difficult if the functions $F_i(\cdot)$ are convex rather than strictly convex. Luckily, an easy and fully distributed procedure can be defined for our cutting-plane consensus algorithm. To derive the reconstruction method, we consider the *linear program dual* of the approximate problem. We will see that this dual linear program has a very favorable structure, which provides additional insights into the proposed method.

3.4.2 Linear Programming Dual Interpretation

Let in the following $\bar{\boldsymbol{y}}^{[i]}(\ell)$ be the solution to (3.16) computed by processor $i\in\mathbf{V}$ at iteration $\ell\in\mathbb{Z}_{>0}$. We will use in the following the short-hand notation $\bar{G}_{i\ell}:=G_i\bar{\boldsymbol{y}}_i(\ell)$ and $\bar{F}_{i\ell}:=F_i(\bar{\boldsymbol{y}}_i(\ell))$. We introduce the index set $L_i\subset\mathbb{Z}_{>0}$, indexing the cutting planes generated by processor $i\in\mathbf{V}$. Recall that the cutting-planes, generated by the proposed constraint generating oracle, take the simple form

$$\nu_i-\bar{G}_{i\ell}^\top\pi\le\bar{F}_{i\ell}, \tag{3.17}$$

and the linear approximate program (2.3) can be written as

$$\begin{aligned}\max_{\pi\ge0,\nu_i}\quad & -\mathbf{h}^\top\pi+\sum_{i=1}^n\nu_i\\ \text{s.t.}\quad & \nu_i-\bar{G}_{i\ell}^\top\pi\le\bar{F}_{i\ell},\quad \ell\in L_i,\ i\in\mathbf{V}.\end{aligned} \tag{3.18}$$

We can now formulate the linear programming dual to the approximate program (2.3). Let $\lambda_{j\ell}\in\mathbb{R}_{\ge0}$, $\ell\in L_i, i\in\mathbf{V}$ be the Lagrange multiplier to the corresponding constraint (3.17). The linear programming dual to (3.18) takes the standard form (2.5). In fact, it is a linear program with the following structure:

$$\begin{aligned}\min_{\lambda_{i\ell}\ge0}\ & \sum_{i=1}^n\sum_{\ell\in L_i}F_i(\bar{\boldsymbol{y}}_i(\ell))\lambda_{i\ell}\\ & \sum_{i=1}^n\sum_{\ell\in L_i}(G_i\bar{\boldsymbol{y}}_i(\ell))\,\lambda_{i\ell}\le\mathbf{h},\\ & \sum_{\ell\in L_i}\lambda_{i\ell}=1,\ i\in\{1,\dots,n\},\end{aligned} \tag{3.19}$$

where we used that $\bar{G}_{i\ell}:=G_i\bar{\boldsymbol{y}}_i(\ell)$ and $\bar{F}_{i\ell}:=F_i(\bar{\boldsymbol{y}}_i(\ell))$. The problem (3.19) optimizes over the convex combinations of the constraints and the objective functions evaluated at the points $\bar{\boldsymbol{y}}^{[i]}(\tau)$. Therefore, the problem (3.19) provides us with an alternative interpretation of the cutting-plane consensus algorithm: *The cutting-plane consensus algorithm corresponds to an inner linearization of the original problem* (3.13). Please note this elegant

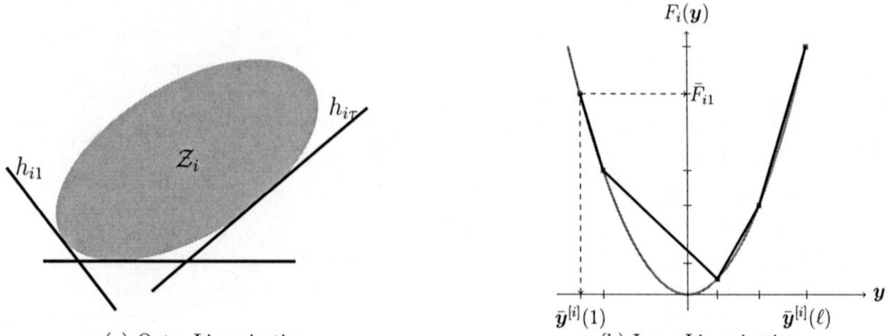

(a) Outer Linearization. (b) Inner Linearization.

Figure 3.1: Outer linearization of the set $\mathcal{Z}_i$ with cutting-planes $h_{i\tau}$ and the inner linearization of a function $F_i(\boldsymbol{y})$ between the points $\bar{\boldsymbol{y}}^{[i]}(\ell)$.

duality relation: while the original cutting-plane consensus algorithm provides an outer linearization of the feasible set of the dual, it provides at the same time an inner linearization of the original problem.

We can use this observation now to provide a reconstruction method for the primal solution. Therefore, we assume now that the algorithm is halted in a way, such that all processors have the same basis. Any processor can now reconstruct its component of the optimal solution. Let $\lambda^*_{i\ell}$ $\ell_i \in L_i, i \in \mathbf{V}$ be the optimal solutions to (3.19). Note that $\lambda^*_{i\ell}$ can be either found by solving explicitly (3.19) or as the dual solution of (3.18). Now, each processor computes its component of the solution as the convex combination of the "linearization points"

$$\boldsymbol{y}_i^* = \sum_{\ell \in L_i} \bar{\boldsymbol{y}}_i(\ell)\lambda^*_{i\ell}.$$

The resulting solution vector is locally feasible, since $\boldsymbol{y}_i^* \in \mathcal{Y}_i$ and $\mathcal{Y}_i$ is convex. Additionally, the complete solution vector $\boldsymbol{y}^* = [\boldsymbol{y}_1^{*\top}, \dots, \boldsymbol{y}_n^{*\top}]^\top$ is globally feasible. To see this, note that

$$\mathbf{h} = \sum_{i=1}^{n} \sum_{\ell \in L_i} (G_i \bar{\boldsymbol{y}}_i(\ell))\, \lambda^*_{i\ell} = \sum_{i=1}^{n} G_i \left(\sum_{\ell \in L_i} \bar{\boldsymbol{y}}_i(\ell)\lambda^*_{i\ell} \right) = \sum_{i=1}^{n} G_i \boldsymbol{y}_i^*.$$

This shows an important property of the cutting-plane consensus algorithm: *A feasible solution to* (3.19) *can always be used to reconstruct a feasible solution, even if the algorithm has not yet converged to the optimal solution.* The next natural question that arises, is whether an optimal solution of the approximate program provides also an optimal solution of the original problem.

Proposition 3.6. *Suppose the cutting-plane consensus algorithm has converged to an optimal solution to* (3.9), *then the recovered primal solution* $\boldsymbol{y}_i^* = \sum_{\ell \in L_i} \bar{\boldsymbol{y}}_i(\ell)\lambda_{i\ell}$ *is an optimal primal solution to* (3.13).

Proof. Strong duality implies that the optimal value of (3.19) is equivalent to the value of the linear approximate problem, which we denote with $F^* = \sum_{i=1}^{n} \sum_{\ell \in L_i} F_i(\bar{\boldsymbol{y}}_i(\ell))\lambda^*_{i\ell}$.

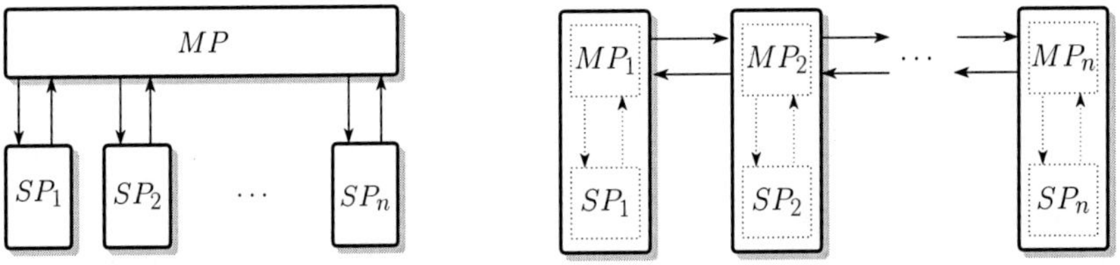

(a) Structure of the classical Dantzig-Wolfe decomposition.

(b) Structure of the cutting-plane consensus algorithm for separable problems.

Figure 3.2: Comparison of the classical master / subproblem structure of the DW decomposition and the peer-to-peer structure of the cutting-plane consensus algorithm.

Jensen's inequality implies now that

$$F^* = \sum_{i=1}^{n} \sum_{\ell \in L_i} F_i(\bar{\boldsymbol{y}}_i(\ell))\lambda_{i\ell}^* \geq \sum_{i=1}^{n} F_i(\sum_{\ell \in L_i} \bar{\boldsymbol{y}}_i(\ell)\lambda_{i\ell}^*) =: \sum_{i=1}^{n} F_i(\boldsymbol{y}_i^*),$$

for all $\sum_{\ell \in L_i} \lambda_{i\ell} = 1$ Since $\boldsymbol{y}^* = [\boldsymbol{y}_1^*, \ldots, \boldsymbol{y}_n^*]$ is a feasible solution it must hold that $\sum_{i=1}^{n} F_i(\boldsymbol{y}_i^*) = F^*$. □

We want to conclude this section by pointing out that the cutting-plane consensus algorithm applied to the almost separable problem (3.13) has strong similarities to the classical *Dantzig Wolfe decomposition* (Dantzig and Wolfe, 1961). The "constraint generation" method is, when considering the dual, a "column generation" method. In the classical Dantzig-Wolfe decomposition one central master program is considered that takes the form (3.19). Repeatedly columns are generated and added to the master program. Exploiting this connection to the Dantzig Wolfe decomposition, we name (3.16) the *local subproblem* SP_i, as it corresponds to the subproblem of the Dantzig Wolfe decomposition used for generating columns. The approximate linear program formed by each processor corresponds to the reduced master program, see Dantzig (1963). However, instead of having one master program, every processor keeps and updates a local version of the master program, i.e., MP_i. We want to point out that a main contribution of our result is that the number of constraints in the master program can be bounded (by known bounds). This is true, since our method is such that inactive or non-basic constraints can be dropped. Only this hard bound on the complexity of the master program makes the method suited for distributed optimization in peer-to-peer processor networks. Recall that only the use of the unique solution to linear programs, i.e., the use of the minimal 2-norm solution, allowed us to drop inactive constraints. A constraint dropping was known to be possible when the underlying problem is linear. However, to the best of our knowledge the method proposed here is the first to solve general convex problems correctly when all inactive constraints are dropped. The difference in the communication structure between the classical decomposition method and the novel cutting-plane consensus algorithm is illustrated in Figure 3.2.

3.4.3 CPC-based Trajectory Exchange Method

We turn our attention now again to the model predictive control problem (3.5). As previously shown, the problem can be formulated as an almost separable convex problem and the cutting-plane consensus algorithm is suitable to solve it. Not surprising, the algorithm has a very characteristic interpretation, when it is applied to solve this model predictive control problem.

Recall that the local constraints of the problem are parameterized by all feasible output trajectories $\boldsymbol{y}_i \in \mathcal{Y}_i$ and the corresponding value of the primal local control problem $F_i(\boldsymbol{y}_i)$ as defined by (3.6). As before, we call the tuple $(\boldsymbol{y}_i, F_i(\boldsymbol{y}_i))$ a *plan* of system i. A basis of the linear approximate program (i.e., (2.3)) in the cutting-plane consensus algorithm becomes now simply a *collection of plans* (i.e., a collection of predicted output trajectories). We allow now that the processors store and exchanges up to $d = n + r \cdot T$ plans. Each processor uses the plans in its memory to form the linear approximate program (2.3). We can use now the interpretation of the approximate program as an inner-linearization to understand the approximate program better. In fact, solving the linear approximate program is equivalent to computing an *optimal convex combination of the plans a processor has in memory*, i.e., a processors solves the linear program

$$\begin{aligned} \min_{\lambda_{i\ell} \geq 0} \quad & \sum_{i=1}^{n} \sum_{\ell \in L_i} \lambda_{i\ell} F_i(\boldsymbol{y}_i(\ell)) \\ \text{s.t.} \quad & \sum_{i=1}^{n} \sum_{\ell \in L_i} \lambda_{i\ell} \boldsymbol{y}_i(\ell) \leq \mathbf{h}, \quad \sum_{\ell \in L_i} \lambda_{i\ell} = 1,\ i \in \{1, \dots, n\}, \end{aligned} \tag{3.20}$$

with a unique dual solution. We will call in the following the plans for which $\lambda_{i\ell} \neq 0$ in the solution of (3.20) the *defining plans.* Clearly, the defining plans correspond to the active constraints.

The second important component of the cutting-plane consensus algorithm is the generation of cutting planes, by solving the local subproblems (3.16). It is not hard to see that the local sub-problem becomes in the model predictive control framework simply an optimal control problem of the form

$$\begin{aligned} \min \quad & \Big(\sum_{\tau=0}^{T-1} f_i(x_i(t+\tau|t), u_i(t+\tau|t)) + \pi^\top(\tau) y_i(t+\tau|t) \Big) + V_i(x(t+T|t)) \\ \text{s.t.} \quad & x_i(t+\tau+1|t) = A_i x_i(t+\tau|t) + B_i u_i(t+\tau|t), \\ & y_i(t+\tau|t) = C_i x_i(t+\tau|t) + D_i u_i(t+\tau|t), \\ & x_i(t|t) = x_i(t),\ x_i(t+\tau|t) \in \mathcal{X}_i,\ u_i(t+\tau|t) \in \mathcal{U}_i \\ & \tau \in \{0, \dots, T-1\}, \end{aligned} \tag{3.21}$$

where $\pi = [\pi^\top(0), \dots, \pi^\top(T-1)]^\top$ is the solution to the linear approximate program (i.e., the dual solution to (3.20)). That is, each system has to solve a local optimal control problem involving a penalty term for the output trajectory defined by the multiplier π. Thus, *the generation of a new cutting-plane corresponds exactly to the computation of a new optimal output trajectory* for the corresponding system.

Therefore, we can provide the following interpretation of the cutting-plane consensus algorithm applied to distributed, finite horizon optimal control problems. First, the processors store, exchange and update plans. Whenever a processors receives a set of plans, it computes the optimal convex combination of all plans. It then checks, whether a new plan, computed by penalizing its own output trajectory by the multiplier vector of the convex combination combination, and eventually adds this plan. It is remarkable, that in contrast to the original trajectory exchange method, the systems exchange up to $d = n + r \cdot T$ trajectories. Thus, each system can alter the plans of the other systems to some extent. Each system chooses the plan for the other systems as a convex combination of their plans. Since the systems are, by assumption, linear, the convex combination will again be a feasible point. Now, since each system can adjust the plans of the other systems, it gets the additional degree of freedom that is necessary to improve the solution and to compute a new, better plan. We also want to emphasize that it is not necessary to run the algorithm until convergence. In contrast, the algorithm can be stopped at any time, providing a feasible solution. Since the single processors exchange sets of feasible plans, they can stop the algorithm after a finite number of communication rounds, find an agreement on the best solution, and consider the optimal convex combination of the plans as the feasible plans. In this sense, the cutting-plane consensus algorithm directly complements the existing trajectory exchange methods.

3.5 Application Example: Distributed Microgrid Control

We discuss next the implementation of the algorithm for a specific control problem. In particular, we present a computational study of the cutting-plane consensus algorithm applied to a *distributed microgrid control* problem. Microgrids are local collections of

Algorithm 3 Dual cutting-plane consensus - Trajectory Exchange Algorithm

`Processor state:` $B^{[i]}(t)$: set of up to $n + r \cdot T$ feasible plans
`Initialization:` $B^{[i]}(0)$: set of n feasible plans

`Function MSG:` **Return** $B^{[i]}(t)$;

`Function STF:` **Set** $Y^{[i]}(t) := \cup_{j \in \mathcal{N}_I(i,t)} \mathtt{MSG}(B^{[j]}(t))$;

(S1) $H^{[i]}_{tmp}(t) \leftarrow B^{[i]}(t) \cup Y^{[i]}(t)$;

(S2) $(\pi, \nu) \leftarrow$ unique dual solution to (3.20) *(compute optimal convex combination)*;
$B^{[i]}_{tmp}(t) \leftarrow$ set of defining plans;

(S3) (i) $(\bar{\boldsymbol{y}}_i, \gamma^*) \leftarrow$ optimal solution and value of (3.21) *(solve optimal control problem)*
(ii) if $\gamma^* < \nu_i$ define $h = (\bar{\boldsymbol{y}}_i, F(\bar{\boldsymbol{y}}_i))$.

(S4) $B^{[i]}(t+1) \leftarrow$ Defining plans of $B^{[i]}_{tmp}(t) \cup h(z^{[i]}(t))$.

distributed energy sources, energy storage devices and controllable loads. Most existing control strategies still use a central controller to optimize the operation (Zamora and Srivastava, 2010). However, for several reasons, detailed, e.g., in Zamora and Srivastava (2010), distributed control strategies, that do not require to collect all data at a central coordinator, are highly desirable. The cutting-plane consensus algorithm provides a tool to realize the microgrid management in a fully distributed way.

We consider the optimization model of a microgrid recently described in Kraning et al. (2012). A microgrid consists of several generators, controllable loads, storage devices and a connection to the main grid. In the following, we use the notational convention that energy generation corresponds to positive variables, while energy consumption corresponds to negative variables. The model we employ here considers the following components of a microgrid.

- A *generator* generates power $p_{gen}(t), t \in [0,T]$ within the absolute bounds $\underline{p}(t) \leq p_{gen}(t) \leq \bar{p}(t)$ and the rate constraints

$$\underline{r}(t) \leq p_{gen}(t+1) - p_{gen}(t) \leq \bar{r}(t).$$

  The stage cost to produce power by a generator is modeled as a quadratic function $f_{gen} = \alpha p_{gen}(t) + \beta p_{gen}^2(t)$.

- A *storage device* can store or release power $p_{st}(t), t \in [0,T]$ within the bounds $-d_{st} \leq p_{st}(t) \leq c_{st}$. The charge level of the storage device is then

$$q_{st}(t+1) = q_{st}(t) + p_{st}(t), \quad q_{st}(0) = q_{st,init}$$

  and must be maintained between $0 \leq q_{st}(t) \leq q_{max}$. Note that $p_{st}(t)$ takes negative values if the storage device is charged and positive values if it is discharged.

- A *controllable load* has a desired load profile $l_{cl}(t)$ and incorporates a stage cost if the load is not satisfied, i.e.,

$$f_{cl}(t) = \alpha(l_{cl}(t) - p_{cl}(t))_+,$$

  where $(z)_+ = \max\{0, z\}$.

- Finally, the microgrid has one *connection to the main grid* that can trade energy. The maximal energy that can be traded is $|p_{tr}| \leq E$. The cost to sell or buy energy is modeled as

$$f_{tr} = -c(t)p_{tr}(t) + \gamma^T |p_{tr}(t)|,$$

  where $c(t)$ the (predicted) price at time t and γ is a general transaction cost.

- The *power demand* $D(t)$ in the microgrid is predicted over a horizon T and should be matched with the generated power at each time instant.

The control objective is to minimize the cost of power generation while satisfying the overall demand. This control problem can be directly represented in the form (3.13), with the local objective functions $f_i = \sum_{t=0}^{T} f_i(t)$, the right-hand side vector of the coupling constraint as the predicted demand $\mathbf{h} = [D(1), \ldots, D(T)]^T$ and $\mathcal{X}_i$ as the local constraints of each unit. The cutting-plane consensus algorithm can solve this problem in a distributed way.

We present simulation results for an example set-up with $n = 101$ decision units, i.e., 60 generators, 20 storage devices, 20 controllable loads and one connection to the main grid. A random demand is predicted for 15 minute time intervals over a horizon of three hours, based on a constant off-set, a sinusoidal growth and a random component. The algorithm is initialized with each processor computing a basis out of the box-constraint set $\{z : -10^5 \cdot \mathbf{1} \leq z \leq 10^5 \cdot \mathbf{1}\}$, leading to a very high initial objective value. Figure 3.3 shows the largest objective value over all processors, relative to the best solution found as the algorithm is continued to perform. The evolution of the objective value over the communication rounds is shown for three different k-circulant graphs. It can be clearly seen that the convergence speed depends strongly on the structure of the communication graph. The convergence for a network with a circular communication structure with two neighbors is significantly slower than for a network with a higher number of neighbors. We also want to emphasize the observation that the difference in the convergence speed between $k = 8$ and $k = 32$ is not as big as the increased communication would let one expect. This shows that the improvement obtained from more communication between the processors becomes smaller with more communication. A good performance of the algorithm can be obtained with little communication between the processors. Please note that for all communication graphs the cutting-plane consensus algorithm requires only few communication rounds to converge to a fairly good solution. Although the convergence to an exact optimal solution might take more iterations, a good sub-optimal solution can be found after very few communication rounds. This property makes the cutting-plane consensus attractive for control and decision applications.

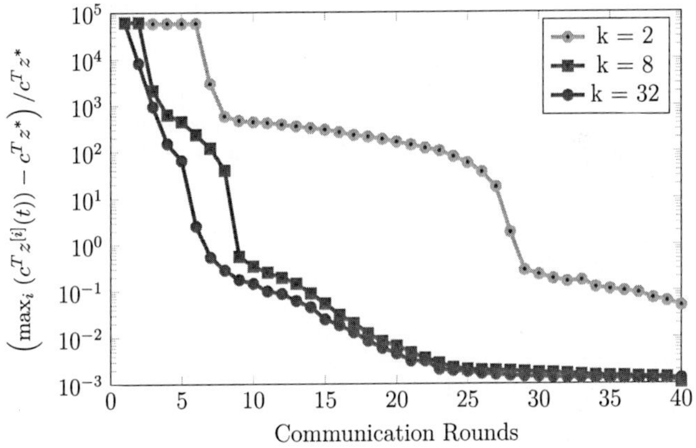

Figure 3.3: Trajectories of the scaled maximal optimal value of the linear approximate programs for different k-circulant communication graphs $\mathcal{G}_c$.

Asynchronous Implementation

The proposed algorithm imposes only very little requirements on the communication network. Although the algorithm has been discussed throughout this thesis for synchronous communication networks, it is easy to see that the algorithm can be easily modified to perform in *asynchronous communication* networks. An algorithm in asynchronous communication networks is such that the different processors do not wait at predetermined times to receive a message, but rather continue with their local computations at their own speed and communicate data irregularly (Bertsekas and Tsitsiklis (1997)). One can easily modify the algorithms proposed here for asynchronous communication networks. In particular, one must only augment each processor with an additional local memory (that can have finite capacity). In this local memory, the cutting-planes (or predicted plans) received from neighboring processors are stored. If the local memory has only a finite capacity, messages received from other processors are written to the memory only if it has enough free capacity and are deleted otherwise.

Each processor repeats then the following three tasks at its own speed: *(A1) it adds all cutting-planes in its local memory to its local optimization problem and voids its local memory; (A2) it performs its local communication, i.e., it computes a new query point, generates a cutting-plane, and updates its basis; (A3) it writes its basis to the local memory of its out-neighbors.*

Due to the local memory, the processors must not synchronize their communication phases, but rather each processor can perform its computations at its own pace. Naturally, this can lead to situations, where a fast processor performs significantly more local computation (and communicates more often) than a slow processor in the same network. However, under very weak assumptions (i.e., joint connectivity of the communication graph) the algorithm is guaranteed to converge.

An *asynchronous implementation* of the algorithm for a microgrid management problem[2] is reported in Lorenzen et al. (2013) and Lorenzen (2013). A fully asynchronous implementation of the algorithm on sixteen different, spatially distributed personal computers, using a UDP communication protocol, was realized. It was shown in Lorenzen et al. (2013) and Lorenzen (2013), that the algorithm performs correctly, although the computational speed of the processors varied significantly, i.e., the fastest processor completed four times as many computational rounds as the slowest processors. Additionally, it was shown that the algorithm is highly tolerant against packet losses. The packet-loss rates do have an influence on the convergence speed of the algorithm, but (under weak assumptions) not on its correctness. The computational studies presented in Lorenzen et al. (2013) and Lorenzen (2013) clearly show that the proposed algorithm is well suited for optimization problems that have to be solved over real world communication networks.

[2]The optimization problem discussed in the named references is similar to the problem considered here, but additionally considers an adjustable control component to react on uncertainties in the demand predictions.

3.6 Conclusions

In the previous two chapters, we have developed a general and broad framework for distributed convex and robust optimization using a polyhedral approximation method. Based on the general problem formulation in Section 2.3, we have proposed an abstract algorithm for distributed optimization in asynchronous communication networks, namely the cutting-plane consensus algorithm. The algorithm is well scalable to large networks in the sense that the amount of data each processor has to store and process is small and independent of the network size. The appealing property of the considered outer approximation method lies in the fact that it imposes very little requirements on the structure of the constraint sets and that it is therefore applicable to a broad variety of decision and control problems. We have shown, how the general algorithm can be specified to various specific problem formulations including semi-definite or robust optimization problems.

However, beyond the natural representations of the optimization problem, we have also shown that the cutting-plane consensus algorithm can be efficiently used in distributed control problems. We have considered a basic finite horizon optimal control problem, including several linear dynamical systems coupled by a constraint. First, we have shown that the problem can be represented as an almost separable optimization problem, if the output trajectories are assumed to be the decision variables. The dual problem of this almost separable problem is a linear program with semi-infinite constraints. By exploiting this problem representation, we were able to see that trajectory exchange methods, as they are well studied in distributed model predictive control, are in fact outer approximation methods for the dual problem. However, it also turns out that the classical trajectory exchange methods are under-parameterized, in the sense that not enough trajectories are stored and exchanged. To fix this issue, we have shown that also the cutting-plane consensus algorithm is applicable to this problem class. On the one hand, the cutting-plane consensus algorithm becomes the dual version of a distributed, asynchronous Dantzig-Wolfe decomposition method. On the other hand, in the context of finite horizon optimal control problems the algorithm becomes a trajectory exchange method. The novel trajectory exchange methods are such that the single systems do not take the trajectories of the other systems as given, but are allowed to vary them as a convex combination of all trajectories in memory. In this way, it can be ensured that the method eventually converges to the optimal solution. We illustrated the applicability and the efficiency of the algorithm on a distributed control problem, namely the economic power dispatch problem in a microgrid. In summary, we presented in this part of the thesis a novel, general framework including several algorithms for distributed decision making and control. The algorithms and the corresponding theoretical derivations provide a basis for the design of distributed systems and explain how an efficient and reliable distributed decision making can be performed.

Up to now, the objective of this thesis was to find constructive tools for decision making in distributed systems. In the last chapter, duality theory took the key role in our analysis. We change this perspective now and consider in the following a reversed situation, where distributed optimization is used for analyzing the behavior of interacting systems and for an optimal cooperative control design. Again, *duality theory* will take the central role in the analysis, and will lead the way to efficient algorithmic solutions.

Chapter 4

Duality and Network Theory in Cooperative Control

4.1 Introduction

We have seen in the previous chapter that *duality* is a concept of major importance in optimization theory. Every optimization problem has a dual problem that admits bounds or even the exact value of the original problem. The duality relation between almost separable convex problems and semi-infinite optimization programs, which was exploited in the previous chapter, is one example, where duality turns out to be extremely elegant and practically relevant.

Another class of optimization problems, that admit a fairly complete exposition on duality theory, are optimization problems over networks, generally known as *network optimization* (Rockafellar, 1998). For network optimization problems a unifying theory based on convex optimization and duality is well-known. The key element of this framework is a pair of two dual optimization problems: the *optimal network flow problem* and the *optimal potential problem*. In optimal flow problems, flows are assigned optimally to the edges of a network, while in optimal potential problems, potentials on the nodes and tensions along the edges are selected in an optimal manner. The duality between these two problem representations has led the way to many important and efficient optimization algorithms (Rockafellar, 1998) (Bertsekas, 1998).

However, duality turns out to be not only important in optimization theory, but also in *controls and systems theory*. For example, the duality between estimation and control is the key element in optimal regulator theory. It comes as no surprise that much of the duality theory found in controls is a result of the important role optimization plays for both the design and analysis of these systems.

A recent trend in modern control theory is the study of cooperative control problems amongst groups of dynamical systems that interact over an information exchange network. A fundamental goal for the analysis of these systems is to reveal the interplay between properties of the individual dynamic agents, the underlying network topology, and the interaction protocols that influence the functionality of the overall system (Mesbahi and Egerstedt, 2010), (Olfati-Saber et al., 2007). Within the numerous control theoretic approaches being pursued to define a general theory for networks of dynamical systems, *passivity* (see Willems (1972)) takes an outstanding role; see e.g., Bai et al. (2011). In Arcak (2007) a passivity based framework for group coordination problems was established. Passivity was used by Zelazo and Mesbahi (2010) to derive performance bounds on the input/output behavior of consensus-type networks. Passivity is also widely used in coordinated control of robotic systems (Chopra and Spong, 2006) and the teleoperation of UAV swarms (Franchi et al., 2011). Passivity-based cooperative control with quantized measurements is studied in De Persis and Jayawardhana (2012). The refined concept of incremental passivity provides a framework to study various synchronization problems (Stan and Sepulchre, 2007), (Scardovi et al., 2010). Passivity was also used in the context of Port-Hamiltonian systems on graphs to establish a unifying framework for a variety of networked dynamical systems in Van der Schaft and Maschke (2013).

The passivity-based cooperative control framework has many modeling similarities to the network optimization framework. We investigate in this chapter the question, whether the cooperative control framework inherits any of the duality results found in network optimization. Somehow surprising, it turns out that convex optimization plays a key role in the analysis of the cooperative behavior of such dynamical networks. We show that the asymptotic behavior of the dynamical networks corresponds exactly to the solutions of a family of four convex optimization problems. Always two of these four optimization problems are Lagrange *duals* to each other. With this perspective, we will establish a novel connection between passivity-based cooperative control and the general network optimization theory of Rockafellar (1998).

The duality relation we present here is not only interesting due to the theoretical insights it provides, but it is also of practical relevance. First, the duality theory will lead us the way to an efficient distributed controller design for inventory systems. Additionally, the duality theory opens the door for analyzing more advanced emergent behaviors of cooperative networks. Just to mention as a preview, we will derive from the duality theory a mechanism to analyze and predict the clustering structure of certain dynamical networks. Ultimately, the duality theory will lead us the way to a computationally attractive method to detect hierarchical clustering structures in certain dynamic networks or in weighted graphs. Thus, in both optimization theory as well as cooperative control the relevance of duality theory cannot be overstated.

The control theoretic framework that will allow us to establish the connection to network theory is the concept of *equilibrium independent passivity* (EIP), see Hines et al. (2011), Jayawardhana et al. (2007). Equilibrium independent passivity generalizes the classical passivity concept to dynamical systems that admit different equilibria for different constant input signals. We consider a general cooperative control framework, where the nodes of the network are EIP systems. As a first contribution of this chapter, we show that constraints on the control inputs, imposed by the network structure of the cooperative control problem, lead to a connection between the control problem and the dual pair of optimization problems, i.e., an optimal flow and an optimal potential problem. With this observation, we provide an interpretation of the inputs and outputs as a pair of dual variables, namely as divergences and potentials, respectively. Next, we consider a class of dynamic feedback control laws, based upon an internal model approach (Wieland et al., 2011), that achieve output agreement. We show that the outputs and the states of the controller can also be interpreted as dual variables, namely as flows and tensions. These results provide us with a complete interpretation of the dynamic variables in a network theoretic sense. We show then that the duality relation between the different variables, and in particular the *conversion formula* relating these variables, leads directly to a Lyapunov function for the closed-loop systems. In this sense, we use convex optimization and duality theory here as a *tool for the analysis of the cooperative behavior of dynamical networks.*

The remainder of the chapter is organized as follows. We prepare the discussions by reviewing some important concepts related to graph and network theory in Section 4.2. As we are considering interacting dynamical systems, the graph theoretic concepts we use in this part are slightly different from those, we considered previously. However, graph theory will only be a supportive concept in this part of the thesis. Building upon the ideas of graph theory, we will introduce the notion of *network theory* following Rockafellar (1998). Inspired by various real world networks such as transportation networks, electrical circuits or communication networks, network theory provides a unifying framework for network analysis and introduces an abstract characterization of concepts such as "flow", "divergence", " potential", and "tension". We review next the conceptual framework here and show its relevance for passivity-based cooperative control problems. Following this, we introduce and discuss the system property equilibrium independent passivity, which is going to be the basis for our later considerations. The passivity based cooperative control framework is then introduced in Section 4.3, where also the important duality relations are derived. We use the duality relations then to construct a Lyapunov function for the dynamical network. Finally, in Section 4.4 an inventory control problem is presented as an exemplary system, where the novel duality relations turn out to be of significant practical relevance.

4.2 Preliminaries

4.2.1 Algebraic Graph Theory

We will use again the concept of a graph $\mathcal{G} = (\mathbf{V}, \mathbf{E})$, with nodes $\mathbf{V} = \{1, \ldots, n\}$ and edges $\mathbf{E} = \{1, \ldots, m\}$ to describe the interaction topology between systems.[1] In contrast to the previous parts of the thesis, we will restrict our attention in the following to undirected graphs. However, it will be convenient to introduce an arbitrary orientation on the edges. We will use in the following sections the elegant result that graphs can be represented algebraically as matrices, see e.g., Godsil and Royle (2001). The matrix representation of a graph, that will turn out to be most important in this thesis, is the directed *incidence matrix.* The incidence matrix $E \in \mathbb{R}^{n \times m}$ of the (directed) graph $\mathcal{G}$ with arbitrary orientation is a $\{0, \pm 1\}$ matrix with the rows and columns indexed by the nodes and edges of $\mathcal{G}$ such that

$$[E]_{ik} = \begin{cases} \text{`+1'} & \text{if node } i \text{ is the initial node of edge } k \\ \text{`-1'} & \text{if node } i \text{ is the terminal node of edge } k \\ 0 & \text{otherwise.} \end{cases} \tag{4.1}$$

As an example, consider the network with five nodes and eight edges illustrated in Figure

[1]We use here the simplified notation to denote the nodes and edges of a network by numbers. It would be more precise to define $\mathbf{V} = \{v_1, \ldots, v_n\}$. However, the simplified notation used here is sufficiently precise for our purposes.

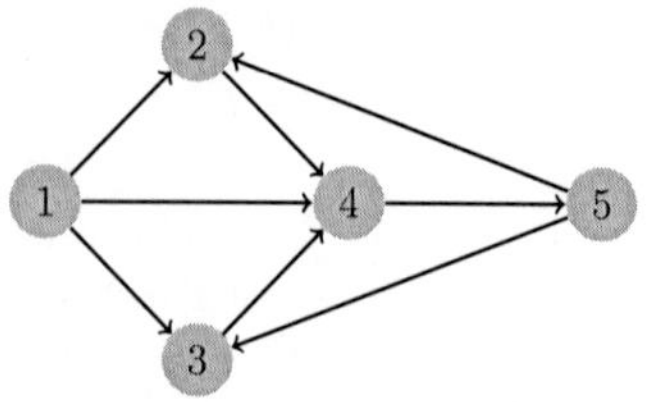

Figure 4.1: Exemplary configuration of a graph with directed edges.

4.1. The incidence matrix for this network is easily determined as

$$E = \begin{bmatrix} 1 & 1 & 1 & 0 & 0 & 0 & 0 & 0 \\ -1 & 0 & 0 & 1 & 0 & 0 & -1 & 0 \\ 0 & -1 & 0 & 0 & 1 & 0 & 0 & -1 \\ 0 & 0 & -1 & -1 & -1 & 1 & 0 & 0 \\ 0 & 0 & 0 & 0 & 0 & -1 & 1 & 1 \end{bmatrix}.$$

The *Laplacian matrix* L of the undirected graph $\mathcal{G}$ can be derived from the incidence matrix as $L = EE^\top$ (Godsil and Royle, 2001). If additional non-negative weights are assigned to the edges, i.e., $q_1, \ldots, q_m$, the weighted graph Laplacian matrix can be computed as $L_Q = EQE^\top$ with $Q = \text{diag}\{q_1, \ldots, q_m\}$.

The definition of the incidence matrix implies directly that any column of E has exactly two nonzero elements, one element '+1' and one elements '-1', and it must obviously hold that for any graph $\mathbf{1}^\top E = 0$. Thus the null-space $\mathcal{N}(E^\top)$ contains the all-ones vector. The following result is well-known in algebraic graph theory (Godsil and Royle, 2001). Let $\mathcal{G}$ be a graph with n nodes and c connected components, then rk $E = n - c$. Thus, the next result follows immediately.

Proposition 4.1. *Let $\mathcal{G}$ be an undirected, connected graph and let E be the directed incidence matrix with arbitrary orientation, then $\mathcal{N}(E^\top) = \text{span}\, \mathbf{1}$.*

We will call $\mathcal{N}(E^\top)$ the *agreement space* of the graph $\mathcal{G}$. From the fundamental theorem of linear algebra (Strang, 1986) follows directly that the agreement space is orthogonal to the column space of the incidence matrix, i.e., $\mathcal{N}(E^\top) \perp \mathcal{R}(E)$.

The row space of the incidence matrix, i.e., $\mathcal{R}(E^\top)$, is called the *differential space* of $\mathcal{G}$ Rockafellar (1998). In fact, given a non-zero vector $\mathbf{y}$ defined in the node space (one might think of $\mathbf{y}$ as a potential), the vector $\zeta = E^\top \mathbf{y}$ characterizes the differences between values (i.e., the potentials) of adjacent nodes. The orthogonal complement to the differential space is the *flow space* $\mathcal{N}(E)$. From linear algebra considerations follows directly, that for a graph $\mathcal{G}$ with n nodes, m edges and c connected components, it holds that dim $\mathcal{N}(E) = m - n + c$. The flow space is closely related to the cycles of $\mathcal{G}$. Consider the following idea of a signed path vector.

Definition 4.2. *A* signed path vector $s \in \mathbb{R}^m$ *of a connected graph $\mathcal{G}$ corresponds to a path such that the i-th element of s takes the value '+1' if edge i is traversed positively, '-1' if traversed negatively, and '0' if the edge is not used in the path.*

A simple *cycle* in a graph is a path where the initial and terminal nodes are the same. Now, the cycles provide an explicit characterization of the flow-space.

Theorem 4.3 (Godsil and Royle (2001)). *Let $\mathcal{G}$ be a connected graph and let E be the directed incidence matrix, then the flow space $\mathcal{N}(E)$ is spanned by all the linearly independent signed path vectors corresponding to the cycles in $\mathcal{G}$.*

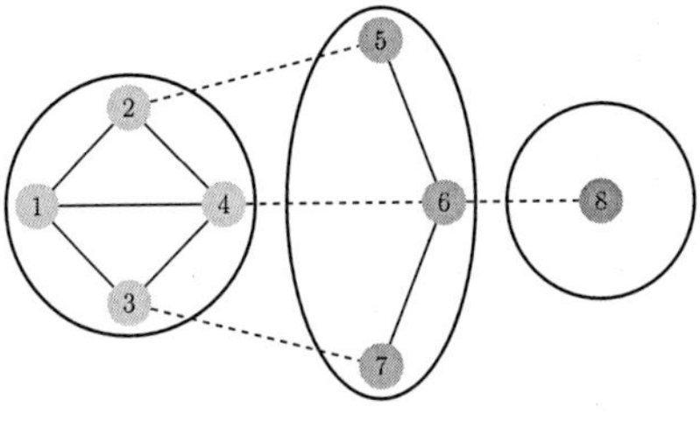

Figure 4.2: Network with eight nodes, partitioned in three clusters.

We will sometimes use the concept of a subgraph of $\mathcal{G}$. A graph $\mathcal{G}' = (\mathbf{V}', \mathbf{E}')$ is a *subgraph* of $\mathcal{G}$ if $\mathbf{V}' \subseteq \mathbf{V}$ and $\mathbf{E}' \subseteq \mathbf{E}$; equivalently, we write $\mathcal{G}' \subseteq \mathcal{G}$. Subgraphs can be induced by either a node set or an edge set. For example, the subgraph $\mathcal{P} \subseteq \mathcal{G}$ induced by the node set $\mathbf{P} \subseteq \mathbf{V}$ is the graph $\mathcal{P} = (\mathbf{P}, \mathbf{E}')$, with $\mathbf{E}' = \{k = (i,j) \,|\, i,j \in \mathbf{P}, k \in \mathbf{E}\}$. Similarly, the subgraph $\mathcal{Q} \subseteq \mathcal{G}$ induced by the edge set $\mathbf{Q} \subseteq \mathbf{E}$ is the graph $\mathcal{Q} = (\mathbf{V}', \mathbf{Q})$, with $\mathbf{V}' \subseteq \mathbf{V}$ the set of all nodes incident to the edges in $\mathbf{Q}$. A disconnected graph can be expressed as the union of connected subgraphs; each connected subgraph is referred to as a *component* of $\mathcal{G}$.[2] This leads now directly to the concept of a cluster.

Definition 4.4. *A* cluster $\mathcal{P}$ *is a connected subgraph of $\mathcal{G}$ induced by a node set $\mathbf{P} \subseteq \mathbf{V}$.*

Definition 4.5. *A* p-partition *of the graph $\mathcal{G}$ is a collection of node sets $\mathbb{P} = \{\mathbf{P}_1, \dots, \mathbf{P}_p\}$ with $\mathbf{P}_i \subseteq \mathbf{V}$, $\cup_{i=1}^{p} \mathbf{P}_i = \mathbf{V}$, and $\mathbf{P}_i \cap \mathbf{P}_j = \emptyset$ for all $\mathbf{P}_i, \mathbf{P}_j \in \mathbb{P}$, such that each subgraph $\mathcal{P}_i$ induced by the node sets $\mathbf{P}_i$ is connected.*

Each subgraph $\mathcal{P}_i$ induced by a p-partition is also a cluster. At times we will also refer to the *p-cluster* of a graph to mean the set of subgraphs induced by a p-partition. For a connected graph $\mathcal{G}$, the union of all clusters induced by a partition will not reconstruct the original graph ($p \geq 2$); that is, $\cup_{i=1}^{p} \mathcal{P}_i \subset \mathcal{G}$. This is formalized by the definition of a *cut-set.*

Definition 4.6. *A* cut-set *of the graph $\mathcal{G}$ is a set of edges whose deletion leads to an increase in the number of connected components in $\mathcal{G}$.*

According to this definition, a cut-set always induces a p-partition ($p \geq 2$). Similarly, any p-partition ($p \geq 2$) of a graph will induce a cut-set. In this case, the cut-set is defined as $\mathbf{Q} = \{(i,j) \in \mathbf{E} | i \in \mathbf{P}_s, j \in \mathbf{P}_t, \forall \mathbf{P}_s, \mathbf{P}_t \in \mathbb{P},\ s \neq t\}$.

Keeping these important concepts of graph theory in mind, we can turn our attention to the more refined idea of *networks.*

[2] We will use from here on the convention that bold-faced capital letters refer to sets, as in $\mathbf{V}$, and the script letters refer to graphs, as in $\mathcal{Q}$.

4.2.2 Network Theory

Network theory generalizes models of different physical processes, such as transportation networks and passive electrical circuits, into one unifying mathematical theory. Based on the underlying physical examples a vocabulary, including the notions of flows and potentials, has been adopted into the description of network optimization problems, see Rockafellar (1998). We rely here on this language as well.

Formally, a *network* is described by a connected graph $\mathcal{G} = (\mathbf{V}, \mathbf{E})$ with an arbitrarily assigned orientation. Please note that all results presented here are independent of the chosen orientation of the graph. The incidence matrix turns out to be important for modeling such physical networks and becomes the key component in a generalized theory.

We call a vector $\boldsymbol{\mu} = [\mu_1, \ldots, \mu_m]^\top \in \mathbb{R}^m$ a *flow* of the network $\mathcal{G}$. An element of this vector, μ_k, is the *flux* of the edge $k \in \mathbf{E}$. An intuitive interpretation for the flux of edge k relates to transportation networks, where it represents the amount of material transported across that edge. The incidence matrix can be used to describe a type of conservation relationship between the flow of the network along the edges and the net in-flow (or out-flow) at each node in the network, termed the *divergence* of the network $\mathcal{G}$. The relationship states that the net flux entering a node must be equal to the net flux leaving the node. The divergence associated with the flow $\boldsymbol{\mu}$ is denoted by the vector $\mathbf{u} = [u_1, \ldots, u_n]^\top \in \mathbb{R}^n$ and can be represented as

$$\mathbf{u} + E\boldsymbol{\mu} = 0. \tag{4.2}$$

In a transportation network, $\mathbf{u}$ can be understood as the "in/outflow" or "leakage" at the network nodes. In an electric circuit interpretation (4.2) corresponds to *Kirchhoff's Current Law.* Note that for any given, fixed divergence vector, the corresponding flow vector is not uniquely defined but can be varied in the circulation space, i.e., $\mathcal{N}(E)$.

The dual concept to flow and divergence relates to potentials and tensions. We simply call a vector defined in the node space $\mathbf{y} \in \mathbb{R}^n$ a *potential* of the network $\mathcal{G}$. An element of this vector, y_i, is called the potential at node i. To any edge $k = (i, j)$ one can associate the potential difference across the edge as $\zeta_k = y_j - y_i$; we also call this the *tension* of the edge k. The tension vector of the network $\mathcal{G}$, $\boldsymbol{\zeta} = [\zeta_1, \ldots, \zeta_m]^\top$, can be expressed as

$$\boldsymbol{\zeta} = E^\top \mathbf{y}. \tag{4.3}$$

From a physical analogy, one might think of the tension to represent a force. In an electric circuit interpretation (4.3) corresponds *Kirchoff's Voltage Law.* Note that for a given and fixed tension vector $\boldsymbol{\zeta}$, the corresponding potential vector is not uniquely defined but can be varied in the agreement space $\mathcal{N}(E^\top)$. Based on the previous discussion, one can easily see that flows and tensions are related to potentials and divergences by the *conversion formula*

$$\boldsymbol{\mu}^\top \boldsymbol{\zeta} = -\mathbf{y}^\top \mathbf{u}. \tag{4.4}$$

Broadly speaking, network theory connects elements of graph theory to a family of convex optimization problems. The beauty of this theory is that it admits elegant and simple duality relations, as it relates a dual pair of optimization problems: the *optimal flow problem* and the *optimal potential problem.*

The first problem we consider attempts to optimize the flow and divergence in a network subject to the conservation constraint (4.2). As an illustrative explanation of this problem, consider a transportation network. The flow $\boldsymbol{\mu}$ represents the goods transported along the lines of the network, while the divergence $\mathbf{u}$ represents the external supply at the nodes. In the most general formulation, a cost is incorporated for both, transporting goods and using external supply. Thus, each edge is assigned a *flux cost* $C_k^{flux}(\mu_k)$ and each node is assigned a *divergence cost* $C_i^{div}(\mathrm{u}_i)$. The optimal flow problem is then

$$\begin{aligned} \min_{\mathbf{u},\boldsymbol{\mu}} \quad & \sum_{i=1}^{n} C_i^{div}(\mathrm{u}_i) + \sum_{k=1}^{m} C_k^{flux}(\mu_k) \\ \text{s.t.} \quad & \mathbf{u} + E\boldsymbol{\mu} = 0. \end{aligned} \tag{4.5}$$

The optimization problem (4.5) admits a dual problem with a very characteristic structure. To form the dual problem, one can replace the divergence u_i and flow μ_k variables in the objective functions with artificial variables $\tilde{\mathrm{u}}_i$ and $\tilde{\mu}_k$, respectively, and introduce the artificial constraints $\mathrm{u}_i = \tilde{\mathrm{u}}_i$, $\mu_k = \tilde{\mu}_k$. These artificial constraints can be dualized with Lagrange multipliers $\mathbf{y} = [\mathrm{y}_1, \dots, \mathrm{y}_n]^\top$ and $\boldsymbol{\zeta} = [\zeta_1, \dots, \zeta_m]^\top$, respectively, leading to the dual problem

$$\begin{aligned} \max_{\mathbf{y},\boldsymbol{\zeta}} \ \min_{\mathbf{u},\tilde{\mathbf{u}},\boldsymbol{\mu},\tilde{\boldsymbol{\mu}}} \quad & \sum_{i=1}^{n} \Big(C_i^{div}(\tilde{\mathrm{u}}_i) + \mathrm{y}_i(\mathrm{u}_i - \tilde{\mathrm{u}}_i) \Big) + \sum_{k=1}^{m} \Big(C_k^{flux}(\tilde{\mu}_k) + \zeta_k(\mu_k - \tilde{\mu}_k) \Big) \\ \text{s.t.} \quad & \mathbf{u} + E\boldsymbol{\mu} = 0. \end{aligned}$$

The latter problem is in fact equivalent to

$$\begin{aligned} \max_{\mathbf{y},\boldsymbol{\zeta}} \ \min_{\mathbf{u},\boldsymbol{\mu}} \quad & -\sum_{i=1}^{n} C_i^{div,\star}(\mathrm{y}_i) - \sum_{k=1}^{m} C_k^{flux,\star}(\zeta_k) + \mathbf{y}^\top \mathbf{u} + \zeta^\top \boldsymbol{\mu} \\ \text{s.t.} \quad & \mathbf{u} + E\boldsymbol{\mu} = 0, \end{aligned}$$

where

$$C_i^{div,\star}(\mathrm{y}_i) = -\inf_{\tilde{\mathrm{u}}_i}\{C_i^{div}(\tilde{\mathrm{u}}_i) - \mathrm{y}_i\tilde{\mathrm{u}}_i\}$$

and

$$C_k^{flux,\star}(\zeta_k) = -\inf_{\tilde{\mu}_k}\{C_k^{flux}(\tilde{\mu}_k) - \zeta_k\tilde{\mu}_k\},$$

i.e., $C_i^{div,\star}(\mathrm{y}_i)$ and $C_k^{flux,\star}(\zeta_k)$ are the convex conjugates of the divergence and flow cost, respectively.[3] Clearly, the minimization part is only finite if $-E^\top \mathbf{y} + \boldsymbol{\zeta} = 0$. Thus, the Lagrange multipliers y_i are node *potentials* and ζ_k are *tensions*, respectively. The dual problem to (4.5) is now called the *optimal potential problem*

$$\begin{aligned} \min_{\mathbf{y},\boldsymbol{\zeta}} \quad & \sum_{i=1}^{n} C_i^{pot}(\mathrm{y}_i) + \sum_{k=1}^{m} C_k^{ten}(\zeta_k), \\ \text{s.t.} \quad & \boldsymbol{\zeta} = E^\top \mathbf{y} \end{aligned} \tag{4.6}$$

with $C_i^{pot}(\mathrm{y}_i) := C_i^{div,\star}(\mathrm{y}_i)$ and $C_k^{ten}(\zeta_k) := C_k^{flux,\star}(\zeta_k)$.

[3] See Appendix A for further explanations.

The two optimization problems (4.5) and (4.6) provide a natural explanation of duality, since both problems admit an interpretation that matches the physical intuition. A similar duality relation will be the basis for the following discussions. We will recover duality results in a certain passivity-based cooperative control framework. Analogous to the classical network theory, the novel duality results we will derive later on, will lead to a novel interpretation of the systems variables. We will provide, for example, an interpretation of the inputs of a control system as divergences and of the outputs as potentials.

4.2.3 Equilibrium Independent Passivity

The main objective of this chapter is to connect a cooperative control problem for a class of passive system to the general network theory introduced in Section 4.2.2. The right systems theoretic framework to establish such a connection turns out to be *equilibrium independent passivity* (EIP), as recently introduced by Hines et al. (2011). Equilibrium independent passivity specializes the classical passivity concept to systems whose equilibrium point depends on an external signal. This framework can be traced back to earlier works, including Jayawardhana et al. (2007). A brief introduction to the classical passivity concept is given in Appendix A2. We introduce here the novel concept of EIP and some complementing results following Hines et al. (2011). For the clarity of presentation, we restrict the discussion here to SISO systems.

Consider a continuous time nonlinear dynamical control system of the form

$$\begin{aligned} \dot{x}(t) &= f(x(t), u(t)), \\ y(t) &= h(x(t), u(t)), \end{aligned} \tag{4.7}$$

with state vector $x(t) \in \mathcal{X} \subseteq \mathbb{R}^p$, and $u(t) \in \mathcal{U} \subseteq \mathbb{R}$, $y(t) \in \mathcal{Y} \subseteq \mathbb{R}$. The system (4.7) is said to have an *equilibrium input-to-state characteristic* if there is a set $\bar{\mathcal{U}} \subseteq \mathcal{U}$ such that for every $\mathrm{u} \in \bar{\mathcal{U}}$ there is a unique $\mathrm{x} \in \mathcal{X}$ satisfying $f(\mathrm{x}, \mathrm{u}) = 0$. The *equilibrium input-to-state map* $k_{\mathrm{x}} : \bar{\mathcal{U}} \mapsto \bar{\mathcal{X}} \subseteq \mathcal{X}$ is defined accordingly by $f(k_{\mathrm{x}}(\mathrm{u}), \mathrm{u}) = 0$. We assume in the following that k_{x} is continuous on $\bar{\mathcal{U}}$. The corresponding *equilibrium input-output map* is defined as $k_{\mathrm{y}} : \bar{\mathcal{U}} \mapsto \bar{\mathcal{Y}} \subset \mathcal{Y}$ with $k_{\mathrm{y}} := h(k_{\mathrm{x}}(\mathrm{u}), \mathrm{u})$. As a notational convention, we will use italic letters for dynamic variables, e.g., $x(t)$, and letters in normal font for constant signals, such as, e.g., x.

Definition 4.7 (Hines et al. (2011)). *The system* (4.7) *is* equilibrium independent passive *(EIP) on* $\bar{\mathcal{U}}$ *if for every* $\mathrm{u} \in \bar{\mathcal{U}}$ *there exists a positive definite* C^1 *storage function* $S_{\mathrm{u}} : \mathcal{X} \mapsto \mathbb{R}_{\geq 0}$ *such that* $S_{\mathrm{u}}(\mathrm{x}) = S_{\mathrm{u}}(k_{\mathrm{x}}(\mathrm{u})) = 0$ *and*

$$\dot{S}_{\mathrm{u}} = \nabla S_{\mathrm{u}}(x(t))^\top f(x(t), u(t)) \leq (u(t) - \mathrm{u})(y(t) - \mathrm{y}), \tag{4.8}$$

where $\mathrm{x} = k_{\mathrm{x}}(\mathrm{u})$ *and* $\mathrm{y} = k_{\mathrm{y}}(\mathrm{u})$.

Note that $S_{\mathrm{u}}(x(t))$ is defined for a given equilibrium input $\mathrm{u} \in \bar{\mathcal{U}}$, and it is required that there exists a storage function $S_{\mathrm{u}}(x(t))$ for all $\mathrm{u} \in \bar{\mathcal{U}}$. We will not distinguish the storage functions defined for different constant input signals in the following and simply write $S(x)$.

Definition 4.8 (Hines et al. (2011)). *The system* (4.7) *is* output strictly equilibrium independent passive *(OSEIP) if*

$$\dot{S} \leq (u(t) - \mathrm{u})(y(t) - \mathrm{y}) - \rho(y(t) - \mathrm{y}) \tag{4.9}$$

for some positive definite function $\rho(\cdot)$.

It was shown in Hines et al. (2011) that the equilibrium input-output map of an OSEIP system is co-coercive. We provide the result here from a slightly different perspective, involving the inverse of the input-output map. The result follows from an observation about co-coercive functions given in Zhu and Marcotte (1995).

Lemma 4.9 (Strong Monotonicity). *If* (4.7) *is OSEIP with* $\rho(y(t) - \mathrm{y}) = \gamma\|y(t) - \mathrm{y}\|^2$ *and* $k_{\mathrm{y}}(\mathrm{u})$ *is invertible on* $\bar{\mathcal{U}}$*, then the inverse* $k_{\mathrm{y}}^{-1}(\mathrm{y})$ *is strongly monotone on* $\bar{\mathcal{Y}}$.

Proof. The dissipation inequality (4.9) with $\rho(y(t) - \mathrm{y}) = \gamma\|y(t) - \mathrm{y}\|^2$ must hold for any trajectory and in particular for any other equilibrium trajectory $u(t) = \mathrm{u}'$ (with $y(t) = \mathrm{y}', x(t) = \mathrm{x}'$ defined accordingly), giving $\nabla S(\mathrm{x}')^\top f(\mathrm{x}', \mathrm{u}') \leq -\gamma\|\mathrm{y}' - \mathrm{y}\|^2 + (\mathrm{y}' - \mathrm{y})(\mathrm{u}' - \mathrm{u})$. Using $f(\mathrm{x}', \mathrm{u}') = 0$ and replacing $\mathrm{u}' = k_{\mathrm{y}}^{-1}(\mathrm{y}')$ $(\mathrm{u} = k_{\mathrm{y}}^{-1}(\mathrm{y}))$ we obtain the strong monotonicity condition $\gamma\|\mathrm{y}' - \mathrm{y}\|^2 \leq (\mathrm{y}' - \mathrm{y})(k_{\mathrm{y}}^{-1}(\mathrm{y}') - k_{\mathrm{y}}^{-1}(\mathrm{y}))$. □

In this work we are interested in systems with "global" input-output maps, having $\bar{\mathcal{U}} = \mathbb{R}$ and $\bar{\mathcal{Y}} = \mathbb{R}$. We can briefly discuss a certain class of systems, for which this property can be established. We focus in the following on systems with constant input and output vector fields of the form

$$\dot{x}(t) = f(x(t)) + gu(t), \; y(t) = h^\top x(t), \tag{4.10}$$

with $f : \mathbb{R}^n \mapsto \mathbb{R}^n$ a C^1 map, and $g, h \in \mathbb{R}^n$. We implicitly assume in the following that neither g nor h are trivially the all zeros vectors.

Proposition 4.10. *Consider the system* (4.10) *and assume (i)* $\lim_{\|x\|\to\infty} \|f(x)\| = \infty$*, (ii)* $\det Df(x) \neq 0, \quad \forall x \in \mathbb{R}^n$*, (iii)* $h^\top(Df(x))^{-1}g \geq \alpha > 0, \forall x \in \mathbb{R}^n$*, where* $Df(x) \in \mathbb{R}^{n\times n}$ *is the Jacobian matrix of* $f(x)$*. Then* $\bar{\mathcal{U}} = \mathbb{R}$*,* $\bar{\mathcal{Y}} = \mathbb{R}$*. Furthermore* $k_{\mathrm{y}} : \mathbb{R} \mapsto \mathbb{R}$ *is invertible and the inverse is continuous and strongly monotone.*

Proof. Assumptions (i) and (ii) imply that $f(x)$ is proper and a local diffeomorphism. From the *global inverse function theorem* Wu and Desoer (1972) follows that $f(x)$ is a C^1 diffeomorphism on $\mathbb{R}^n$ to $\mathbb{R}^n$. Thus, f^{-1} exists, is a C^1 function, and for any $\mathrm{u} \in \mathbb{R}$ there is a unique $\mathrm{x} \in \mathbb{R}^n$ such that $\mathrm{x} = f^{-1}(g\mathrm{u})$. We have now $k_{\mathrm{y}}(\mathrm{u}) = h^\top f^{-1}(g\mathrm{u})$. This is a continuous function. Additionally, we have for any $\mathrm{u}, \mathrm{u}' \in \mathbb{R}$, that

$$(\mathrm{u}' - \mathrm{u})(k_{\mathrm{y}}(\mathrm{u}') - k_{\mathrm{y}}(\mathrm{u})) = (\mathrm{u}' - \mathrm{u})h^\top(f^{-1}(g\mathrm{u}') - f^{-1}(g\mathrm{u})).$$

Consider now the real valued function $\ell(\mathrm{u}) = h^\top f^{-1}(g\mathrm{u})$. By the mean value theorem, we conclude that $\ell(\mathrm{u}') - \ell(\mathrm{u}) = \nabla\ell(s)(\mathrm{u}' - \mathrm{u})$, where $s = \lambda g\mathrm{u}' + (1 - \lambda)g\mathrm{u}$, for some $\lambda \in [0, 1]$. Clearly, this is equivalent to $h^\top f^{-1}(g\mathrm{u}') - h^\top f^{-1}(g\mathrm{u}) = \left(h^\top Df^{-1}(s)g\right)(\mathrm{u}' - \mathrm{u})$. Since $f(x)$ is a diffeomorphism, we have $Df^{-1}(s) = \left(Df(\hat{s})\right)^{-1}|_{\hat{s}=f^{-1}(s)}$. Therefore, we can conclude strong monotonicity from

$$(\mathrm{u}' - \mathrm{u})(k_{\mathrm{y}}(\mathrm{u}') - k_{\mathrm{y}}(\mathrm{u})) = (\mathrm{u}' - \mathrm{u})\left(h^\top\left(Df(\hat{s})\right)^{-1} g\right)(\mathrm{u}' - \mathrm{u}) \geq \alpha(\mathrm{u}' - \mathrm{u})^2.$$

□

Having established conditions for the existence of a globally defined input-output map, we can provide a condition for OSEIP for the systems (4.10). The following result is a variation of a result provided in Jayawardhana et al. (2007).

Proposition 4.11. *Assume the system* (4.10) *is passive (in the classical sense) with a twice continuously differentiable storage functions* $S(x(t)) : \mathcal{X} \mapsto \mathbb{R}_{\geq 0}$ *and there is some* $\delta > 0$ *such that*

$$[f(x(t)) - f(\bar{\mathrm{x}})]^\top [\nabla S(x(t)) - \nabla S(\bar{\mathrm{x}})] \leq -\delta \|x(t) - \bar{\mathrm{x}}\|^2 \tag{4.11}$$

for all $\mathrm{x} \in \bar{\mathcal{X}}$. *Then the system* (4.10) *is OSEIP with the equilibrium independent storage function* $S_{\mathrm{u}}(x(t))$,

$$S_{\mathrm{u}}(x(t)) = S(x(t)) - x(t)^\top \nabla S(\bar{\mathrm{x}}) - [S(\bar{\mathrm{x}}) - (\bar{\mathrm{x}})^\top \nabla S(\bar{\mathrm{x}})]. \tag{4.12}$$

Proof. Following the argumentation of Jayawardhana et al. (2007), one can use the nonlinear version of the Kalman-Yakubovich-Popov (KYP) lemma (Khalil, 2002), to prove OSEIP. By the passivity of (4.10) follows that $\nabla S(x(t))^\top g = x^\top(t) h$. Condition (4.11) implies

$$\begin{aligned} &[f(x(t)) - f(\bar{\mathrm{x}})]^\top [\nabla S(x(t)) - \nabla S(\bar{\mathrm{x}})] \leq -\delta \|x(t) - \bar{\mathrm{x}}\|^2 \\ &\leq -\frac{\delta}{h^\top h}(x(t) - \bar{\mathrm{x}})^\top h h^\top (x(t) - \bar{\mathrm{x}}) = -\gamma \|y(t) - \bar{\mathrm{y}}\|^2 \end{aligned}$$

with $\gamma = \frac{\delta}{h^\top h}$. The derivative of the new storage function S_0 satisfies

$$\begin{aligned} \dot{S}_{\mathrm{u}} &= [\nabla S(x(t)) - \nabla S(\bar{\mathrm{x}})]^\top [f(x(t)) + g u(t)] \\ &= [\nabla S(x(t)) - \nabla S(\bar{\mathrm{x}})]^\top [f(x(t)) - f(\bar{\mathrm{x}}) + g u(t) - g \bar{\mathrm{u}}] \\ &\leq -\delta \|x(t) - \bar{x}\|^2 + (x(t) - \bar{x})^\top h (u(t) - \bar{\mathrm{u}}) \\ &\leq -\gamma \|y(t) - \bar{\mathrm{y}}\|^2 + (y(t) - \bar{\mathrm{y}})^\top (u(t) - \bar{\mathrm{u}}). \end{aligned}$$

The proof of the positive definiteness of S_{u} exploits convexity properties and can be found in Jayawardhana et al. (2007). □

We want to point out two important system classes that admit globally defined equilibrium input-output maps and are OSEIP. We are particularly interested in systems with $k_{\mathrm{y}}(0) \neq 0$, which motivates the presentation of the following examples.

Example 4.12 (Scalar Systems). *Consider the scalar nonlinear system*

$$\dot{x}(t) = -f(x(t)) + u(t), \; y(t) = x(t), \tag{4.13}$$

with $x(t) \in \mathbb{R}, u(t) \in \mathbb{R}, y(t) \in \mathbb{R}$. *The system* (4.13) *is OSEIP if for all* $x'(t)$ *and* $x''(t)$ *it holds that*

$$(x'(t) - x''(t))\big(f(x'(t)) - f(x''(t))\big) \geq \gamma (x'(t) - x''(t))^2,$$

for some constant $\gamma > 0$. *Similar systems are considered in the context of synchronization, for example in Scardovi et al. (2010), DeLellis et al. (2011), or Liu et al. (2011). Note that the QUAD condition reduces for the scalar system* (4.13) *to a strong monotonicity condition on* $f(x(t))$. *The dynamics can then be understood as the gradient of a strongly convex function, i.e.,* $f(x(t)) = \nabla F(x(t))$. *The equilibrium input-output map is the strongly monotone function* $k_{\mathrm{y}}(\mathrm{u}) = f^{-1}(\mathrm{u})$.

Example 4.13 (Affine Systems). *Consider the affine system of the form*

$$\begin{aligned} \dot{x}(t) &= Ax(t) + Bu(t) + \mathrm{w} \\ y(t) &= Cx(t) + Du(t) + \mathrm{v} \end{aligned}$$

with (A, B) *controllable and* (A, C) *observable. The system is OSEIP if* (A, B, C, D) *satisfy the matrix equations of the KYP-lemma, (i.e., the system is strictly output passive in the classical sense) and* A *is invertible (Hines et al., 2011). In addition, with* $\mathrm{w} \in \mathbb{R}^p$ *and* $\mathrm{v} \in \mathbb{R}$ *being constant signals the equilibrium input-output map is the affine function*

$$k_y(\mathrm{u}) = \left(-CA^{-1}B + D\right)\mathrm{u} + \left(-CA^{-1}\mathrm{w} + \mathrm{v}\right), \tag{4.14}$$

i.e., the dc-gain *of the linear system plus the constant value determined by the exogenous inputs.*

Remark 4.14. *Example* (4.13) *provides an explicit connection to the classical internal model control design Francis (1976), Isidori and Byrnes (1990).*[4] *For simplicity, assume in the following* $\mathrm{v} = 0$. *Consider* w *as a disturbance generated by an exosystem of the form* $\dot{w} = Sw$. *Clearly for the constant disturbance* $w = \mathrm{w}$ *the dynamics can be trivially chosen with* $S = 0$. *Now, there exist* $x^w = \Pi w$ *and* $u^w = \Gamma w$ *that exactly attenuate the disturbance if and only if there exist* Π, Γ *satisfying the Sylvester equation*

$$\Pi S = A\Pi + B\Gamma + I \tag{4.15}$$

Note that (4.15) *is the first part of the classical Francis equations Francis (1976). For the constant disturbance signal, we have* $S = 0$. *Assume additionally that* A *is invertible. The matrix* Π *that satisfies (4.15) for any* Γ *is* $\Pi = -A^{-1}(B\Gamma_i + I)$. *From this, the optimal steady state trajectoriy computes as* $x^w = -A^{-1}(B\Gamma + I)\mathrm{w}$, *and the steady state output as* $y^w = C\Pi\mathrm{w} + D\Gamma\mathrm{w} = -CA^{-1}(B\Gamma + I)\mathrm{w} + D\Gamma\mathrm{w} = (-CA^{-1}B + D)u^w - CA^{-1}\mathrm{w}$. *This clearly coincides with the equilibrium input-output maps* $k_y(u)$.

4.3 Duality in Passivity-based Cooperative Control

We turn our attention now to a multi-agent cooperative control problem. In this section, we will explore the *output agreement* problem for a team of agents that are each passive systems. We connect this dynamical control problem to the network theory framework of Rockafellar (1998). The main contribution of this part is the exposition of several duality relations between the involved dynamical variables.

We consider a network $\mathcal{G} = (\mathbf{V}, \mathbf{E})$ of dynamical systems, with each node of $\mathcal{G}$ representing an *output strictly equilibrium independent passive SISO system*

$$\Sigma_i : \quad \begin{aligned} \dot{x}_i(t) &= f_i(x_i(t), u_i(t)) \\ y_i(t) &= h_i(x_i(t), u_i(t)), \end{aligned} \qquad i \in \mathbf{V}. \tag{4.16}$$

The OSEIP property implies that for each $\mathrm{u}_i \in \bar{\mathcal{U}}_i$ a storage function exists, and we assume that the storage functions $S_i(x_i(t))$ are all positive definite. We focus on networks comprised

[4]This connection was pointed out by Prof. C. De Persis and is reported in Bürger and De Persis (2013).

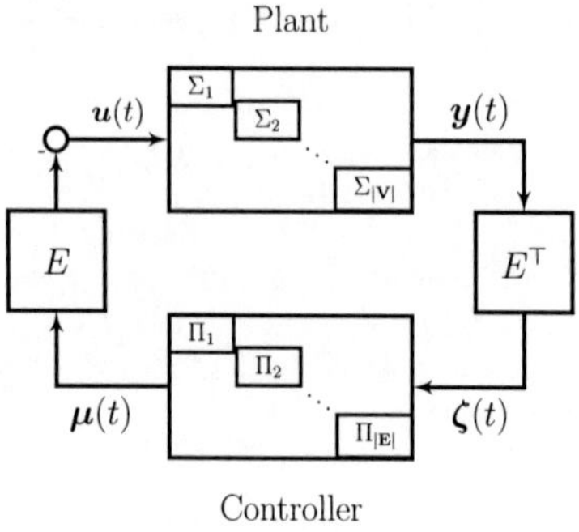

Figure 4.3: Block-diagram of the canonical passivity-based cooperative control structure.

of heterogeneous systems, and in particular on networks where the uncoupled systems have inherently *different unforced equilibria*. Equivalently, each agent has a different equilibrium input-output map when there is no forcing term, i.e.,

$$k_{y,i}(0) \neq k_{y,j}(0) \text{ for all } i, j \in \mathbf{V}. \tag{4.17}$$

In the following we adopt the notation $\boldsymbol{y}(t) = [y_1(t), \ldots, y_n(t)]^\top$ and $\boldsymbol{u}(t) = [u_1(t), \ldots, u_n(t)]^\top$ for the stacked output and input vectors of the complete network. We use normal font letters $\mathbf{y}, \mathbf{u}$ to indicate that a vector corresponds to equilibrium trajectories. Similarly, we use $\mathbf{k}_\mathrm{y}(\mathbf{u}) = [k_{\mathrm{y},1}(\mathrm{u}_1), \ldots, k_{\mathrm{y},n}(\mathrm{u}_n)]^\top$. If all $k_{\mathrm{y},i}$ are invertible, we write $\mathbf{k}_\mathrm{y}^{-1}(\mathbf{y}) = [k_{\mathrm{y},1}^{-1}(\mathrm{y}_1), \ldots, k_{\mathrm{y},n}^{-1}(\mathrm{y}_n)]^\top$.

The control objective is to reach agreement on a constant steady state value of the outputs.

Definition 4.15. *A network of dynamical systems* (4.16) *is said to reach* output agreement *if*

$$\lim_{t\to\infty} \boldsymbol{y}(t) \to \beta \mathbf{1} \tag{4.18}$$

for some $\beta \in \mathbb{R}$, *called the* agreement value.

Output agreement should be achieved through a coupling of the network nodes using the control inputs. The cooperative control framework we consider is based on a canonical control structure, illustrated in Figure 4.3, see e.g., Arcak (2007) or Van der Schaft and Maschke (2013). The controllers Π_k are located between neighboring systems in the network and use the relative outputs of the systems i.e.,

$$\boldsymbol{\zeta}(t) = E^\top \boldsymbol{y}(t),$$

as their inputs. This immediately implies that the couplings between neighboring systems are *diffusive*, see Hale (1997), Wieland (2010). As we consider the underlying graph to be undirected, we focus on symmetric couplings, where the output of a controller influences the two incident systems with reversed signs. Following the network interpretation, the control input $\boldsymbol{u}(t)$ is generated by the mapping of the controller outputs, denoted by $\boldsymbol{\mu}(t)$, through the incidence matrix E to the nodes of the system.

We now make the following observation resulting from the network structure of the control system. The control input $\boldsymbol{u}(t)$ is always in the column space of the incidence matrix, that is

$$\boldsymbol{u}(t) \in \mathcal{R}(E). \tag{4.19}$$

This structural condition is independent of the exact controller we choose, but is purely a consequence of the networked structure of Figure 4.3. A direct consequence of this structural condition is that the control inputs summed over all nodes in the network will vanish, i.e., $\mathbf{1}^\top \boldsymbol{u}(t) = 0$. Before we propose a feedback control law for output agreement, we first investigate the role of the structural constraint $\boldsymbol{u}(t) \in \mathcal{R}(E)$ in the agreement problem. This also motivates the general organization of this section, whereby we first consider the network implications for output agreement at the *plant level*, and then perform a similar analysis at the *control level*, and conclude by analyzing the complete closed-loop system.

4.3.1 The Plant Level

Without specifying a control law, we first investigate the output agreement problem considering only the plant dynamics (4.16) and the structural constraint on the input (4.19). This discussion will already provide us the first duality relation in passivity-based cooperative control.

We note directly that the dynamical network (4.16) has an agreement steady state $\mathbf{y} = \beta\mathbf{1}$ if and only if there exists an equilibrium control input $\mathbf{u} \in \mathcal{R}(E)$ such that $\mathbf{k}_{\mathrm{y}}(\mathbf{u}) \in \mathrm{span}\{\mathbf{1}\}$. The algebraic constraint $\mathbf{u} \in \mathcal{R}(E)$ contributes to the structure of the coupling controller, while the second condition ensures the existence of a steady state output in agreement, since $\mathbf{y} = \mathbf{k}_{\mathrm{y}}(\mathbf{u})$. Note that $\mathrm{span}\{\mathbf{1}\} = \mathcal{N}(E^\top)$. A stronger, more insightful, characterization of the agreement steady state can be established under some additional assumptions.

Assumption 4.1. *All node dynamics* (4.16) *are OSEIP with* $\bar{\mathcal{U}} = \mathbb{R}$ *and* $\bar{\mathcal{Y}} = \mathbb{R}$ *and the equilibrium input-output maps* $k_{\mathrm{y},i}(\mathrm{u}_i)$ *of all systems* (4.16) *are invertible and strongly monotone.*

The affine systems of Example 4.13 and the gradient systems of Example 4.12 satisfy this assumption. Under this assumption the agreement steady state can be precisely characterized.

Lemma 4.16. *Let Assumption 4.1 hold. Then there is a unique agreement steady state* $\mathbf{y} = \beta\mathbf{1}$ *and the agreement value* β *satisfies*

$$\sum_{i=1}^{n} k_{\mathrm{y},i}^{-1}(\beta) = 0.$$

Proof. The condition of a steady state is $\mathbf{k}_{\mathrm{y}}(\mathbf{u}) = \beta\mathbf{1}$. Since, by assumption, all input-output characteristics are invertible, we have $\mathbf{u} = \mathbf{k}_{\mathrm{y}}^{-1}(\beta\mathbf{1})$. Since $\mathbf{u} \in \mathcal{R}(E)$, it must hold that $\mathbf{u}$ is orthogonal to $\mathcal{N}(E^\top)$. Since $\mathcal{N}(E^\top) = \mathrm{span}\{\mathbf{1}\}$, it follows that $\mathbf{1}^\top\mathbf{u} = \mathbf{1}^\top\mathbf{k}_{\mathrm{y}}^{-1}(\beta\mathbf{1}) = 0$. By assumption, the input-output maps are strongly monotone implying that there always exists exactly one β such that $\mathbf{1}^\top\mathbf{k}_{\mathrm{y}}^{-1}(\beta\mathbf{1}) = \sum_{i=1}^{n} k_{\mathrm{y},i}^{-1}(\beta) = 0$. □

This small observation leads us to a novel interpretation of the output agreement problem in the context of network theory. We define for each node the integral function of the equilibrium input-output map $k_{\mathrm{y},i}(\mathrm{u}_i)$, denoted $K_i(\mathrm{u}_i)$, and satisfying

$$\nabla_{\mathrm{u}_i} K_i(\mathrm{u}_i) = k_{\mathrm{y},i}(\mathrm{u}_i). \tag{4.20}$$

As $k_{\mathrm{y},i}$ is co-coercive, the integral function K_i is convex. We will call $K_i(\mathrm{u}_i)$ in the following the *cost function* of node i. The convex conjugate of the cost function $K_i(\mathrm{u}_i)$ is

$$K_i^\star(\mathrm{y}_i) = \sup_{\mathrm{u}_i} \{\mathrm{y}_i \mathrm{u}_i - K_i(\mathrm{u}_i)\}, \tag{4.21}$$

and is called the *potential function* of node i. Recall that $\nabla_{\mathrm{y}} K_i^\star(\mathrm{y}_i) = k_{\mathrm{y},i}^{-1}(\mathrm{y}_i)$. In a next step, we connect the steady state agreement output of the dynamical network with a distributed, symmetric feedback, i.e., $\boldsymbol{u}(t) \in \mathcal{R}(E)$, to a dual pair of network optimization problems.

Optimal Potential Problem: Consider the static *optimal potential problem* of the form

$$\begin{aligned} \min_{\mathrm{y}_i} \; & \sum_{i=1}^{n} K_i^\star(\mathrm{y}_i), \\ \text{s.t.} \; & E^\top \mathbf{y} = 0. \end{aligned} \tag{OPP1}$$

The objective functions of this problem are the convex conjugates of the integral functions of the equilibrium input-output maps. The constraint $E^\top \mathbf{y} = 0$ enforces a balancing of the potentials, i.e., $\mathrm{y}_1 = \cdots = \mathrm{y}_n$. Note that the problem can be written in the standard form of an optimal potential problem (4.6), with $C_i^{pot}(\mathrm{y}_i) := K_i^\star(\mathrm{y}_i)$ and $C^{ten}(\zeta_k)$ the indicator function $I_0(\zeta_k)$ for the point zero. In fact, we have now established that the steady state agreement output of the network of dynamical systems (4.16) corresponds to the optimal potentials computed with (OPP1).

Theorem 4.17. *Let Assumptions 4.1 hold, and let* $\mathbf{y} = \beta\mathbf{1}$ *be the agreement state of the network* (4.16). *Then* $\mathbf{y}$ *is the optimal solution to* (OPP1).

Proof. Under the stated assumptions, Lemma 4.16 shows that an agreement steady state exists with $\mathbf{y} = \beta\mathbf{1}$ and that the agreement value β satisfies $\sum_{i=1}^n k_{\mathrm{y},i}^{-1}(\beta) = 0$. In particular, we have

$$\sum_{i=1}^{n} \nabla K_i^\star(\beta) = \sum_{i=1}^{n} k_{\mathrm{y},i}^{-1}(\beta) = 0.$$

This is the first-order optimality condition of (OPP1) after replacing the constraint $E^\top \mathbf{y} = 0$ with the equivalent condition $\mathbf{y} = \beta\mathbf{1}$. Therefore, $\mathbf{y} = \beta\mathbf{1}$ is an optimal solution to (OPP1). □

This observation leads directly to an interpretation of the systems outputs $\boldsymbol{y}(t)$ as *potentials* in a network theoretic sense. Somehow surprising, the steady state control input generating this agreement output is directly related to the dual problem of (OPP1).

Optimal Flow Problem: The dual problem to (OPP1) can be derived in the standard way as outlined in Section 4.2.2. One obtains the dual cost functions as the convex conjugates of the original cost functions, i.e., $K_i^{\star\star} := K_i$ and $I_0^\star = 0$, leading directly to the *optimal flow problem*

$$\begin{aligned} \min_{\mathbf{u},\boldsymbol{\mu}} \quad & \sum_{i=1}^{n} K_i(\mathrm{u}_i) \\ \text{s.t.} \quad & \mathbf{u} + E\boldsymbol{\mu} = 0, \end{aligned} \tag{OFP1}$$

with divergence variables $\mathbf{u} \in \mathbb{R}^n$ and flow variables $\boldsymbol{\mu} \in \mathbb{R}^m$. This is, in fact, an optimal flow problem of the structure given in (4.5). Please note that $\mathbf{u}$ refers at this point only to the divergence variable of the network optimization problem. However, as suggested by the notation, we will connect the divergence variable to the control inputs. A first observation in this direction is that the equality constraint $\mathbf{u}+E\boldsymbol{\mu} = 0$ incorporates directly the structural condition $\mathbf{u} \in \mathcal{R}(E)$. In fact, the solution to (OFP1) is exactly the steady state input required for output agreement.

Theorem 4.18. *Let Assumptions 4.1 hold and let* $\mathbf{y} = \beta\mathbf{1}$ *be the agreement state of the network* (4.16)*. Then, the unique* $\mathbf{u}$ *satisfying* $\mathbf{k}_\mathrm{y}(\mathbf{u}) = \mathbf{y}$ *is an optimal solution to* (OFP1).

Proof. Consider the Karush-Kuhn-Tucker conditions of optimality for (OFP1), with $\mathbf{y}$ being the Lagrange multiplier to the equality constraint, i.e., $\nabla K_i(\mathbf{u}) - \mathbf{y} = 0$, $\mathbf{y}^\top E = 0$, and $\mathbf{u} + E\boldsymbol{\mu} = 0$. The first condition is equivalent to $\mathbf{k}_\mathrm{y}(\mathbf{u}) = \mathbf{y}$ and the second condition implies $\mathbf{y} = \alpha\mathbf{1}$. Thus, combining the second and the third condition, we obtain $\mathbf{y}^\top\mathbf{u} = 0$, i.e., $\alpha\sum_{i=1}^n \mathrm{u}_i = 0$. This is equivalent to $\sum_{i=1}^n k_{\mathrm{y},i}^{-1}(\mathrm{y}_i) = 0$. Thus, the Lagrange multiplier $\mathbf{y}$ is the agreement steady state output. Additionally, the corresponding steady state input $\mathbf{u}$ satisfies the KKT condition $\mathbf{k}_\mathrm{y}(\mathbf{u}) = \mathbf{y}$, and is therefore an optimal solution to (OFP1). □

Thus, we can interpret the input of the dynamical network $\boldsymbol{u}(t)$ as a *divergence* in the network theoretic sense. Additionally, we have established an interpretation of the outputs $\boldsymbol{y}(t)$ as *potential* variables. As the divergence and potential variables are a pair of primal/dual variables, this observation provides us with a first duality relation between the inputs $\boldsymbol{u}(t)$ and the outputs $\boldsymbol{y}(t)$ of the dynamical network (4.16). In this context, we find that this duality is actually precisely explained via the two dual optimization problems listed above. This duality interpretation can be extended further when considering the conversion formula (4.4) that states $\boldsymbol{\mu}^\top\boldsymbol{\zeta} = -\mathbf{y}^\top\mathbf{u}$. For the problem (OPP1), the edge tensions are fixed to be zero, which in turn forces the solution to its special structure in $\mathrm{span}\{\mathbf{1}\}$. The conversion formula then immediately yields that $\mathbf{y}^\top\mathbf{u} = 0$. Please note that the conversion formula holds in particular also for the dynamic variables. It is straight forward to see that $\boldsymbol{\mu}^\top(t)\boldsymbol{\zeta}(t) = -\boldsymbol{y}^\top(t)\boldsymbol{u}(t)$. However, the connection to the network optimization problems holds only for the asymptotic behavior of the dynamic variables.

We have analyzed the properties of the agreement steady state solution for the dynamical network (4.16) under the structural condition on the control input, i.e., $\boldsymbol{u}(t) \in \mathcal{R}(E)$. While this discussion provides a clear interpretation of the potential and tension variables, as well as of the divergence variables, it leaves some freedom in the definition of the flow variables. In fact, it is remarkable that the edge flow variables $\boldsymbol{\mu}$ in (OFP1) are neither constrained nor penalized by any cost function. Problem (OFP1) does not determine the flows uniquely, as any feasible flow $\boldsymbol{\mu}$ can be varied in $\mathcal{N}(E)$, i.e., the *circulation space* of $\mathcal{G}$, see Rockafellar (1998), Godsil and Royle (2001). If $\mathcal{G}$ contains cycles, then there is a continuum of edge flows $\boldsymbol{\mu}$, corresponding to a given $\mathbf{u}$. We will call in the following all $\boldsymbol{\mu}$ which correspond to an optimal solution of (OFP1) the *feasible flows*. However, the conservation constraint $\mathbf{u} = -E\boldsymbol{\mu}$ indicates already that the flow variables are important for the computation of the steady state control input. In fact, the flows will be determined by the *control structure*, which is chosen to regulate the network to output agreement.

4.3.2 The Control Level

We close the loop now and present a distributed control scheme, that ensures the dynamical network reaches output agreement. We focus therefore on controls in the generic structure illustrated in Figure 4.3 and define the dynamical systems Π_k.

To begin, note that any static diffusive coupling, using only the current relative measurement, i.e., $\big(y_i(t) - y_j(t)\big)$, vanishes as an agreement state is reached. Consequently, for systems with different unforced equilibria, i.e., condition (4.17), output synchronization cannot be achieved using only static couplings. As a direct consequence of the *internal model for synchronization* (Wieland et al., 2011), (Wieland and Allgöwer, 2010), we consider dynamic couplings and use an integrator as common internal model for the entire network with the form

$$\Pi_k : \quad \begin{array}{l} \eta_k(t) := \int_{t_0}^{t}\big(y_j(\tau) - y_i(\tau)\big)d\tau, \\ \mu_k(t) = \psi_k(\eta_k(t)) \end{array} \qquad k \in \mathbf{E}, \tag{4.22}$$

where the edge k is incident to the nodes i and j. The function $\psi_k : \mathbb{R} \mapsto \mathbb{R}$ is a coupling nonlinearity that is, similar to Arcak (2007), assumed to be the gradient of some convex, differentiable function $P_k : \mathbb{R} \mapsto \mathbb{R}_{\geq 0}$, attaining a minimum at the origin, i.e.,

$$\psi_k(\eta_k(t)) := \nabla P_k(\eta_k(t)). \tag{4.23}$$

The control input applied to the plants is computed according to the structure of Figure 4.3 as

$$\boldsymbol{u}(t) = -E\boldsymbol{\mu}(t).$$

In summary, the dynamic coupling control law can be represented as

$$\begin{aligned} \dot{\boldsymbol{\eta}}(t) &= E^\top \boldsymbol{y}(t), \quad \boldsymbol{\eta}(t_0) \in \mathcal{R}(E^\top), \\ \boldsymbol{u}(t) &= -E\boldsymbol{\psi}(\boldsymbol{\eta}(t)), \end{aligned} \tag{4.24}$$

where $\boldsymbol{\eta}(t) = [\eta_1(t), \ldots, \eta_m(t)]^\top$ is the controller state and the controller output is denoted as $\boldsymbol{\psi}(\boldsymbol{\eta}(t)) = [\psi_1(\eta_1(t)), \ldots, \psi_m(\eta_m(t))]^\top$. Please note that $\boldsymbol{\eta}(t) \in \mathcal{R}(E^\top)$ and $\boldsymbol{u}(t) \in \mathcal{R}(E)$ for all times. We leave here some degree of freedom on the nonlinear output functions of the controller. However, for the remainder of this section we will restrict our attention to the following class of functions.

Assumption 4.2. *The functions $P_k(\cdot)$ are twice differentiable, even, and strongly convex on $\mathbb{R}$ for all $k \in \mathbf{E}$.*

To start the analysis of the coupling controller (4.24), we first have to show that it is able to generate the steady state control input, i.e., $\mathbf{u}$, which leads to complete output agreement. Recall that the steady state control input leading to output agreement is the solution to (OFP1). We exploit an interpretation of the controller variables in a network theory framework to show that the internal model controller (4.24) can generate the correct steady state input.

Optimal Potential Problem: We assume in the following that $\mathbf{u} = [\mathrm{u}_1, \ldots, \mathrm{u}_n]^\top$ is a solution to (OFP1). Consider now the following optimization problem

$$\begin{aligned} \min_{\boldsymbol{\eta}, \mathbf{v}} \quad & \sum_{i=1}^{n} \mathrm{u}_i \mathrm{v}_i + \sum_{k=1}^{m} P_k(\eta_k), \\ \text{s.t.} \quad & \boldsymbol{\eta} = E^\top \mathbf{v}. \end{aligned} \tag{OPP2}$$

Problem (OPP2) is in the structure of the optimal potential problem given in (4.6). The variables $\mathbf{v} = [\mathrm{v}_1, \ldots, \mathrm{v}_n]^\top$ are node potentials and are associated to a linear cost function. The variables $\boldsymbol{\eta} = [\eta_1, \ldots, \eta_m]^\top$ are tensions along the edges and are associated to the objective functions P_k, which are the integral functions of the coupling nonlinearities. The next result shows that the optimal solution to (OPP2) directly defines the steady state value of the internal variable $\boldsymbol{\eta}(t)$ of the dynamic controller (4.24), which is necessary for output agreement.

Theorem 4.19. *Let Assumptions 4.2 hold and let* $\mathbf{u}$ *be the solution to* (OFP1). *Then, the solution* $(\mathbf{v}, \boldsymbol{\eta})$ *to* (OPP2) *is such that* $\boldsymbol{\eta} \in \mathcal{R}(E^\top)$ *and* $\mathbf{u} = -E\boldsymbol{\psi}(\boldsymbol{\eta})$.

Proof. We use the following short-hand notation $\mathbf{P}(\boldsymbol{\eta}) = \sum_{k=1}^{m} P_k(\eta_k)$. By assumption, all P_k are strongly convex functions and are globally defined. Thus, the optimization problem (OPP2) has a finite optimal solution. From the Lagrangian

$$\mathcal{L}(\boldsymbol{\eta}, \mathbf{v}, \boldsymbol{\mu}) = \mathbf{P}(\boldsymbol{\eta}) + \mathbf{u}^\top \mathbf{v} + \boldsymbol{\mu}^\top(-\boldsymbol{\eta} + E^\top \mathbf{v}), \tag{4.25}$$

with Lagrangian-multiplier $\boldsymbol{\mu} \in \mathbb{R}^m$, the optimality conditions can be derived as $\nabla \mathbf{P}(\boldsymbol{\eta}) - \boldsymbol{\mu} = 0$, $E^\top \boldsymbol{\mu} + \mathbf{u} = 0$, and $\boldsymbol{\eta} = E^\top \mathbf{v}$. From the last condition follows that any optimal $\boldsymbol{\eta}$ satisfies $\boldsymbol{\eta} \in \mathcal{R}(E^\top)$. Furthermore, the first two conditions can be combined, providing $\mathbf{u} = -E\nabla \mathbf{P}(\boldsymbol{\eta}) := -E\boldsymbol{\psi}(\boldsymbol{\eta})$. □

Please note that although (OPP2) is an optimal potential problem, it is not directly connected to the problem (OPP1), which was associated to the outputs of the dynamical system. Instead, (OPP2) is an additional optimal potential problem, which describes the steady state behavior of the internal model controller (4.24). We conclude with the observation that the internal model controller (4.24) can, in fact, generate the desired steady state output.

Naturally, the dual problem to the potential problem (OPP2) is again a flow problem. This dual problem is related to the output of the internal model controller, and will provide us with the missing interpretation of the network flow variables.

Optimal Flow Problem: The dual problem to (OPP2) can be derived directly from the Lagrangian (4.25). Standard calculations lead to the following optimal flow problem

$$\begin{aligned} \min_{\boldsymbol{\mu}} \quad & \sum_{k=1}^{m} P_k^\star(\mu_k) \\ \text{s.t.} \quad & \mathbf{u} + E\boldsymbol{\mu} = 0. \end{aligned} \tag{OFP2}$$

Please note that in this problem, $\mathbf{u}$ is not a decision variable, but the solution previously defined by (OFP1). However, the problem is in compliance with the standard form of optimal flow problems (4.5), as one can simply introduce artificial divergence variables and add as a cost function the indicator function for the point $\mathbf{u}$.

Note that any flow $\boldsymbol{\mu}$ which is feasible for (OFP2) is also a feasible flow for our first optimal flow problem (OFP1), since $\mathbf{u}$ is by assumption an optimal solution to (OFP1). We can see therefore, that the new flow problem (OFP2) selects from all feasible flows the one that minimizes the flow cost, defined by the convex conjugates of the integral functions of the coupling nonlinearities.

To complete the connection between the dynamic variables of the control systems and the network theory variables, we now formalize the connection between the optimal flow and the output of the internal model controller.

Theorem 4.20. *Let Assumptions 4.2 hold, and let* $\boldsymbol{\eta}$ *be an optimal solution to* (OPP2). *Then the optimal solution* $\boldsymbol{\mu}$ *to* (OFP2) *satisfies* $\boldsymbol{\mu} = \boldsymbol{\psi}(\boldsymbol{\eta})$.

Proof. We use again the short-hand notation $\mathbf{P}^{\star}(\boldsymbol{\mu}) = \sum_{k=1}^{m} P_k^{\star}(\mu_k)$. Consider the Lagrangian of (OFP2)

$$\mathcal{L}(\boldsymbol{\mu}, \mathbf{v}) = \mathbf{P}^{\star}(\boldsymbol{\mu}) + \mathbf{v}^{\top}(-\mathbf{u} - E\boldsymbol{\mu})$$

with Lagrange multiplier $\mathbf{v}$. The conditions of optimality are $\nabla_{\boldsymbol{\mu}} \mathbf{P}^{\star}(\boldsymbol{\mu}) = E^{\top}\mathbf{v}$ and $\mathbf{u} + E\boldsymbol{\mu} = 0$. As by assumption all the functions P_k are strongly convex, their gradients are invertible, and $\nabla P_k^{-1} = \nabla P_k^{\star}$. Thus, the first optimality condition implies that $\boldsymbol{\mu} = \nabla \mathbf{P}(E^{\top}\mathbf{v})$. We can define now $\boldsymbol{\eta} = E^{\top}\mathbf{v}$. To complete the proof, it remains to show that $\boldsymbol{\eta}$ is an optimal solution to (OPP2). We have from the second optimality condition that $\mathbf{u} + E\nabla \mathbf{P}(\boldsymbol{\eta}) = 0$. Thus, $\boldsymbol{\eta}$ satisfies all optimality conditions of (OPP2), which proves the theorem. □

We have now a second duality relation in the passivity-based cooperative control framework. The internal state of the controller (4.24), $\boldsymbol{\eta}(t)$, can be understood as a *tension*, while the output of the controller, $\boldsymbol{\mu}(t)$, can be understood as the corresponding dual *flow* variable.

We want to emphasize again that we indentified network optimization problems on two different levels of the cooperative control problem. First, at the plant level, the dual problems (OPP1) and (OFP1) characterize the properties of the agreement state. The problems are fully determined by the properties of the dynamical systems (4.16) and the topology of the network $\mathcal{G}$. Second, another dual pair of network optimization problems, i.e., (OPP2) and (OFP2), is associated to the internal model controller (4.24), used to achieve output synchronization. These problems depend on the chosen control structure, i.e., the coupling nonlinearities, as well as on the solution to (OFP1). Thus, the two levels are not completely independent, but the plant level problems influence the problems on the control level.

4.3.3 The Closed-Loop Perspective

It remains to analyze the behavior of the closed-loop dynamical system. We exploit the conversion formula (4.4), i.e., $\boldsymbol{\mu}^{\top}\boldsymbol{\zeta} = -\mathbf{y}^{\top}\mathbf{u}$, to construct a Lyapunov function. We first note that the conversion formula also holds for the dynamic variables, i.e., $\boldsymbol{\mu}^\top(t)\boldsymbol{\zeta}(t) = -\boldsymbol{y}^\top(t)\boldsymbol{u}(t)$. The right hand side of this dynamic conversion formula is reminiscent of a supply function for passive dynamical systems. A natural question to ask is what the conversion formula looks like for the supply function $(\boldsymbol{y}(t)-\mathrm{y})^\top(\boldsymbol{u}(t)-\mathrm{u})$. Exploiting the previously established connections, we make the following considerations

$$\begin{aligned}(\boldsymbol{y}(t)-\mathrm{y})^\top(\boldsymbol{u}(t)-\mathrm{u}) &= -\boldsymbol{y}^\top(t)E(\boldsymbol{\mu}(t)-\upmu) - \mathrm{y}^\top(\boldsymbol{u}(t)-\mathrm{u}) \\ &= -\boldsymbol{\zeta}^\top(t)(\boldsymbol{\mu}(t)-\upmu) - \mathrm{y}^\top(\boldsymbol{u}(t)-\mathrm{u}) \\ &= -\dot{\boldsymbol{\eta}}^\top(t)\Big(\nabla\mathbf{P}(\boldsymbol{\eta}(t)) - \nabla\mathbf{P}(\upeta)\Big) - \mathrm{y}^\top(\boldsymbol{u}(t)-\mathrm{u}).\end{aligned} \tag{4.26}$$

Observe that in this particular problem we have $\mathrm{y}\in\mathrm{span}\{\mathbf{1}\}$ and $\boldsymbol{u}(t),\mathrm{u}\in\mathcal{R}(E)$. Therefore, $\mathrm{y}^\top(\boldsymbol{u}(t)-\mathrm{u})=0$. We conclude that

$$\Big(\nabla\mathbf{P}(\boldsymbol{\eta}(t)) - \nabla\mathbf{P}(\upeta)\Big)^\top\dot{\boldsymbol{\eta}}(t) = -(\boldsymbol{y}(t)-\mathrm{y})^\top(\boldsymbol{u}(t)-\mathrm{u}). \tag{4.27}$$

The last equation has the flavor of a dissipation equality. In fact, a storage function corresponding to (4.27) is

$$\mathbf{B}_P(\boldsymbol{\eta}(t),\upeta) = \mathbf{P}(\boldsymbol{\eta}(t)) - \mathbf{P}(\upeta) - \nabla\mathbf{P}(\upeta)^\top(\boldsymbol{\eta}(t)-\upeta). \tag{4.28}$$

Note that (4.28) is the *Bregman distance*, see Bregman (1967), associated with $\mathbf{P}$ between $\boldsymbol{\eta}(t)$ and $\upeta$. The function $B_P(\boldsymbol{\eta}(t),\upeta)$ is positive definite and radially unbounded since $P(\cdot)$ is a strictly convex function. From (4.27) follows now

$$\dot{\mathbf{B}}_P(\boldsymbol{\eta}(t),\upeta) = -(\boldsymbol{y}(t)-\mathrm{y})^\top(\boldsymbol{u}(t)-\mathrm{u}). \tag{4.29}$$

We can use these observations now for a Lyapunov analysis of the closed-loop system.

Theorem 4.21. *Consider the network of dynamical systems* (4.16)*, with the control inputs defined in* (4.24)*. Let Assumption 4.1 and Assumption 4.2 hold. Then the network* (4.16)*,* (4.22) *converges to the agreement steady state* $\mathrm{y}=\beta\mathbf{1}$*, and*

$$V(\boldsymbol{x}(t),\boldsymbol{\eta}(t)) = \mathbf{S}(\boldsymbol{x}(t)) + \mathbf{B}_P(\boldsymbol{\eta}(t),\upeta) \tag{4.30}$$

with $\mathbf{S}(\boldsymbol{x}(t)) := \sum_{i=1}^n S_i(x_i(t))$ *(i.e. the EIP storage functions), is a Lyapunov function for the closed-loop system.*

Proof. The Lyapunov function $V(\boldsymbol{x}(t),\boldsymbol{\eta}(t))$ is positive definite since both $S_i(x_i(t))$ and $\mathbf{B}_P(\boldsymbol{\eta}(t),\upeta)$ are positive definite. By assumption, all systems (4.16) are OSEIP and thus

$$\dot{\mathbf{S}}(\boldsymbol{x}(t)) \le -\gamma\|\boldsymbol{y}(t)-\mathrm{y}\|^2 + (\boldsymbol{y}(t)-\mathrm{y})^\top(\boldsymbol{u}(t)-\mathrm{u}). \tag{4.31}$$

Thus, combining (4.31) and (4.29), we obtain the directional derivative of the Lyapunov function candidate as

$$\dot{V}(\boldsymbol{x}(t),\boldsymbol{\eta}(t)) = \dot{\mathbf{S}}(\boldsymbol{x}(t)) + \dot{\mathbf{B}}_P(\boldsymbol{\eta}(t),\upeta) \le -\gamma\|\boldsymbol{y}(t)-\mathrm{y}\|^2. \tag{4.32}$$

Since $\lim_{t\to\infty} V(\mathbf{x}(t)) - V(\mathbf{x}(t_0))$ exists and is finite, and since ρ_i are positive definite, we conclude from Barbalat's lemma (Khalil (2002)) that $\lim_{t\to\infty}\|y_i(t)-\mathrm{y}_i\| = \lim_{t\to\infty}\|y_i(t)-\beta\|\to 0$. Additionally, by the invertability of the input-output map follows that $\boldsymbol{u}(t)$ converges to u and, consequently, $\boldsymbol{\eta}(t)$ converges to $\upeta$. □

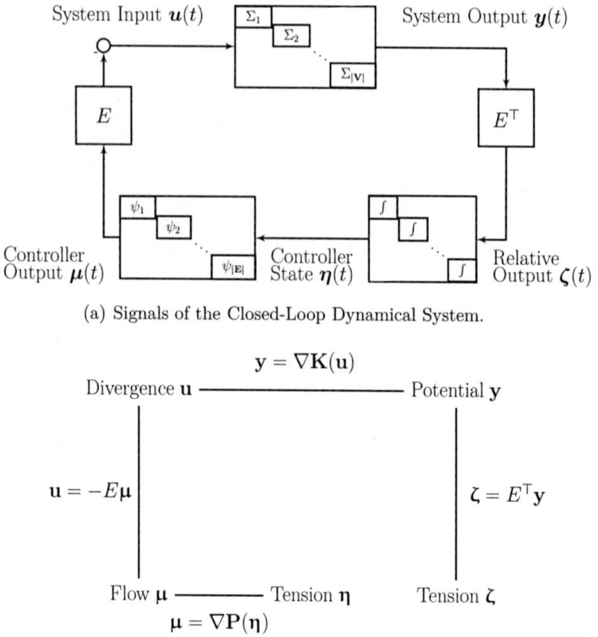

(a) Signals of the Closed-Loop Dynamical System.

(b) Variables of the Network Theoretic Framework.

Figure 4.4: The block diagram of the closed loop system (a) and the abstracted illustration of the network variables (b). Note that the static tension variables $\boldsymbol{\zeta}$ and $\boldsymbol{\eta}$ are not connected, while their dynamic counterparts $\boldsymbol{\zeta}(t)$ and $\boldsymbol{\eta}(t)$ are.

We can summarize the observations of this section as follows. All signals of the dynamical network (4.16) and (4.24) have static counterparts in the network theory framework of Section 4.2.2. For example, the static counterparts of the outputs $\boldsymbol{y}(t)$ are the optimal solutions $\mathbf{y}$ of an optimal potential problem (OPP1), defined by the equilibrium input-output map of the dynamical systems. Equivalently, the corresponding dual variables, i.e., divergence variables in (OFP1), $\mathbf{u}$, are the static counterparts to the control inputs $\boldsymbol{u}(t)$. Additionally, the state $\boldsymbol{\eta}(t)$ and the output $\boldsymbol{\mu}(t)$ of the internal model based controller (4.24) have the tension and flow variables of (OPP2) and (OFP2), respectively, as their static counterparts. We visualize the connection between the dynamic variables of the closed-loop system and the static network variables in Figure 4.4. Note that the signals in the dynamical system influence each other in a closed-loop structure, while there is no equivalent closed-loop relation for the network variables. The two tension variables $\boldsymbol{\zeta}$ and $\boldsymbol{\eta}$ are not connected, while their dynamic counterparts $\boldsymbol{\zeta}(t)$ and $\boldsymbol{\eta}(t)$ are connected by an integrator. Additionally, a summary of all variables involved in the output agreement problem together with their static counterparts and the defining network optimization problems is provided in Table 4.1. For the sake of completeness, we include also the dynamic variable $\boldsymbol{v}(t)$, which corresponds to the potential variables $\mathbf{v}$ of (OPP2). Although we did not consider $\boldsymbol{v}(t)$ explicitly up to now, we can define it in accordance to (OPP2) as $\boldsymbol{\eta}(t) = E^\top \boldsymbol{v}(t)$, and obtain a complete connection between the passivity based control problem and the network optimization problems (OPP1), (OFP1), (OPP2), and (OFP2).

4.4 Application Example: Optimal Distribution Control

As an application example for the novel theory, we consider here the flow control problem in a multi-inventory system. The contribution of this section is the presentation of the results in the context of the network optimization problems discussed above. Routing control in multi-inventory systems has also been studied in Bauso et al. (2006), De Persis (2013), and Wei and van der Schaft (2013).

The control problem is to achieve an optimal routing in a distribution network composed of several inventories with a deteriorating storage, see e.g. Goyal and Giri (2001) for a justification of this model. Consider n inventories with inventory levels $I_i(t)$. The inventory level is influenced by the external supply or demand to this inventory $D_i(t)$, the amount of goods shipped to/from another inventory R_i and a decay rate $\theta_i > 0$, modeling the perishing of goods in the inventory (Goyal and Giri, 2001). This leads to the dynamics of one inventory as

$$\frac{dI_i}{dt} = D_i(t) + R_i(t) - \theta_i I_i(t), \quad i = 1, \dots, n. \tag{4.33}$$

We suppose here that the external demand or supplies at one node are constant, i.e., $D_i(t) = D_i$ and that it is balanced over the network, i.e., $\sum_{i=1}^n D_i = 0$. Goods can be shipped between inventories along m transportation lines of the network. Let E be the incidence matrix, describing the incidence relation of transportation lines and inventories, then we have

$$\mathbf{R}(t) = E\boldsymbol{\mu}, \tag{4.34}$$

where $\mathbf{R} = [R_1, \dots, R_n]^\top$ and $\boldsymbol{\mu}(t)$ is the amount of good transported in the network. A schematic illustration of such an inventory system with 10 inventories is shown in Figure 4.5.

Dynamic Signal		Network Variable		Relation	Cost Function	Problem
$\boldsymbol{y}(t)$	system output	y	potential	$\mathrm{y} = \mathrm{k_y}(\mathrm{u})$	$K_i^\star(\mathrm{y}_i)$	OPP1
$\boldsymbol{\zeta}(t)$	relative output	ζ	tension	$\zeta = E^\top \mathrm{y}$	$I_0(\zeta_k)$	OPP1
$\boldsymbol{u}(t)$	system input	u	divergence	$\mathrm{u} = \mathrm{k_y^{-1}}(\mathrm{y})$	$K_i(\mathrm{u}_i)$	OFP1
$\boldsymbol{\mu}(t)$	controller output	μ	flow	$\mathrm{u} + E\mu = 0$	$P_k^\star(\mu_k)$	OFP2
$\boldsymbol{v}(t)$	–	v	potential	$\eta = E^\top \mathrm{v}$	$\mathrm{u}_k \mathrm{v}_k$	OPP2
$\boldsymbol{\eta}(t)$	controller state	η	tension	$\mu = \psi(\eta)$	$P_k(\eta_k)$	OPP2

Table 4.1: Relation between variables involved in the dynamical system and their static counterparts.

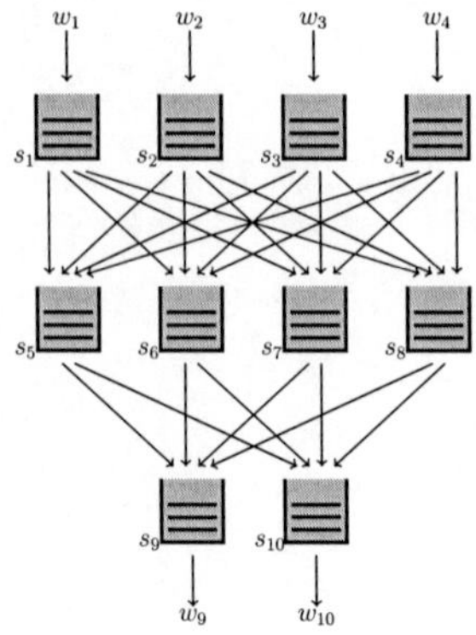

Figure 4.5: Schematic illustration of the multi-inventory system with 10 inventories. The inventories $i = 1, \dots, 4$ are supplied with the amount of goods $w_i > 0$, while there is a constant demand at nodes $i = 9, 10$, with $w_i < 0$.

We assume now additionally that transporting goods within the network incorporates a costs, characterized by the convex function

$$\mathcal{F}_k(\mu_k), \quad k = 1, \dots, m. \tag{4.35}$$

It is reasonable to assume that the transportation capacity of one line is limited. Therefore, we might constrain the flows on one line to the set $\mathcal{W}_k = \{\mu_k : -w_k \leq \mu_k \leq w\}$ for some capacity bound $w_k > 0$. This constraint can be integrated into the flow cost functions by simply defining $\mathcal{F}_k$ such that

$$\mathcal{F}_k(\mu_k) \begin{cases} < \infty & \text{if } \mu_k \in \mathcal{W}_k \\ = +\infty & \text{if } \mu_k \notin \mathcal{W}_k. \end{cases}$$

However, we assume in the following that $\mathcal{F}_k$ is twice continuously differentiable and $\nabla^2 \mathcal{F}_k(\mu_k) > 0$ for all $\mu_k \in \mathcal{W}_k$.

We assume in the following that the optimal routing problem is feasible, that is there exists $\boldsymbol{\mu} \in \text{int}(\mathcal{W}_1) \times \cdots \times \text{int}(\mathcal{W}_m)$, with $\text{int}(\mathcal{W}_k)$ being the interior of Γ_k, such that

$$\mathbf{D} + E\boldsymbol{\mu} = 0,$$

where $\mathbf{D} = [D_1, \dots, D_n]^\top$.

We aim now to design a *distributed control law*, which takes only the imbalance between neighboring inventory systems as inputs and regulates the inventory system to a steady state configuration where the following two objectives are satisfies: (i) all storage levels are balanced (see e.g. De Persis (2013)), and (ii) the flow minimizes the cost induced by the cost functions (4.35). The network theoretic interpretation of the passivity-based cooperative control system presented above, will provide us directly with a solution to this control problem. First, we note that the dynamics of each inventory system (4.33) is equilibrium independent passive if we take the input $u_i(t) = R_i(t)$ and the output $y_i(t) = I_i(t)$. The equilibrium input-output map is then $k_{y,i}(\mathrm{u}_i) = \frac{1}{\theta_i}(\mathrm{u}_i + D_i)$. The optimization problem (OFP1), which is used to determine the equilibrium inputs is now easily formulated as

$$\min_{\mathrm{u}\in\mathcal{R}(E)} \sum_{i=1}^{n} \Big(\frac{1}{2\theta_i} \mathrm{u}_i^2 + \frac{1}{\theta} D_i \mathrm{u}_i \Big).$$

The dual problem (OPP1) can also be easily derived as

$$\min_{\mathrm{y}\in\mathcal{N}(E^\top)} \sum_{i=1}^{n} \Big(\frac{\theta}{2} \mathrm{y}_i^2 - D_i \mathrm{y}_i \Big)$$

From the latter follows directly that the agreement steady state output is $\mathbf{y} = 0$, i.e., the all zeros vector. The agreement steady state input is then $\mathbf{u} = \mathbf{D}$. We aim to solve with the feedback controller the following optimal routing problem

$$\begin{aligned} \min_{\mu} \sum_{k=1}^{m} \mathcal{F}_k(\boldsymbol{\mu}_k) \\ D + E\mu = 0, \end{aligned} \tag{4.36}$$

without knowing the demand/supply vector. Rather, we use only the inventory levels as measurable quantities. We can compare now the optimal routing problem to the problem (OFP2) and see directly, that the two problems are identical with $P_k^\star(\boldsymbol{\mu}_k) = \mathcal{F}_k(\boldsymbol{\mu})$. Thus, a direct consequence of our previous discussion is that the distributed feedback controller

$$\begin{aligned} \dot{\boldsymbol{\eta}}(t) &= E^\top \boldsymbol{I} \\ \boldsymbol{R} &= -E \nabla \boldsymbol{\mathcal{F}}^{-1}(\boldsymbol{\eta}(t)) \end{aligned} \tag{4.37}$$

with $\boldsymbol{I} = [I_1, \ldots, I_n]^T$, and $\boldsymbol{\mathcal{F}}^{-1}(\boldsymbol{\eta}(t)) = \sum_{k=1}^{m} \mathcal{F}_k^{-1}(z_k(t))$, solves the optimal routing problem. Note that the proof of Theorem (4.21) remains valid, since under the stated feasibility assumption of the steady state flow the Bregman distance $\mathbf{B}_{\mathcal{F}^\star}$ is a positive definite function and can serve as a Lyapunov function.

4.5 Conclusions

We have considered in this chapter a cooperative control problem between passive dynamical systems. The dynamical systems were assumed to be equilibrium independent passive systems, that is, the systems were assumed to be passive with respect to any possible equilibrium configuration. To ensure output agreement of the systems, we considered a certain nonlinear dynamic diffusive coupling between the single systems. In a first step, we characterized the possible output configurations on which an agreement can be reached as the solution of a certain optimal potential problem. The potential functions turned out to be the integral functions of the equilibrium input-output configuration of the node dynamical systems. Dual to this potential problem, an optimal flow problem was identified that determines the steady state input, required to achieve output agreement. In a second step, we considered the dynamic controller and identified again a dual pair of network optimization problems, defining the steady state behavior of the controller. The controller output could be associated to an optimal flow problem, while the controller state was connected to an optimal tension variable. In this way, a fairly complete picture of the duality relations in passivity-based cooperative control was established.

Besides these theoretical insights, the optimization oriented network analysis has also several practical implications. We have presented an inventory control problem, where the proposed duality relations lead the way to a distributed controller design. In the next chapter, we will see that the duality relations open the way for the analysis of further, more complex emergent behaviors of interacting dynamical networks. In certain dynamical networks, one can observe the behavior that groups of nodes in the network reach an agreement, while being not in agreement with other nodes. We call this phenomenon clustering or partial output agreement. Somehow surprising, it turns out that the optimization and duality tools proposed here are the right tools to explain such a clustering phenomenon, as we will show in the following chapter.

Chapter 5

Clustering in Dynamical Networks

5.1 Introduction

Clustering, or cluster synchronization, is the emergent phenomenon that in a network of dynamical systems, the network partitions into several groups and all systems within the same group agree upon a common state. To illustrate clustering in dynamical networks visually, the trajectories of two clustering dynamical networks are shown in Figure 5.1. We present the trajectories here only as a motivating preview and will explain the details of the dynamical model later on. At this point, we only want to emphasize that the network nodes form groups (two on the left, six on the right) and reach output agreement only within these groups. The phenomenon is observed across diverse fields ranging from the brain sciences (Escalona-Morán et al., 2007) to social networks (Lazer et al., 2009).. Clustering is intimately related to output agreement, as studied in the previous chapter, and the importance of clustering can be seen best in this context. For example, the synchronization of power networks is crucial for its stability, while frequency clustering in the network can lead to catastrophic failures. It is interesting to understand the behavior of dynamical networks, in the case where complete output agreement is not achieved, but rather a clustering appears.

For complete output agreement, we have already gained a fairly good understanding in the previous chapter, where we could build upon a significant amount of previous work and various control theoretic contributions to understand the underlying mechanisms and requirements. In contrast, the existing literature on clustering is by far not as vast. For example, Wu et al. (2009) and Ma et al. (2006) studied how networks can be forced to cluster according to predefined structures using pinning control or adaptive interaction weights. Xia and Cao (2011) studied different mechanisms leading to clustering in diffusively coupled networks, including structured dynamics, delays, and negative couplings. Some other models are proposed that exhibit clustering inherently by their dynamical properties without having the clustering structure specified *a priori*. One of the most celebrated clustering models is the "bounded confidence opinion dynamics" model (Blondel et al., 2009), where clustering is caused by a state-dependent communication graph. Although the model itself is fairly simple, there exists currently no theory that allows one to predict the resulting clustering structure. In De Smet and Aeyels (2009) and Aeyels and De Smet (2011) a clustering model is presented, where the partitioning of the network is caused by different driving forces applied to the agents along with saturated interaction rules. The development of the aforementioned models is driven by an increasing interest in understanding the mechanisms leading to clustering. In several applications it is also of interest to detect, which parts of the network are strongly connected (with respect to the dynamic behavior)

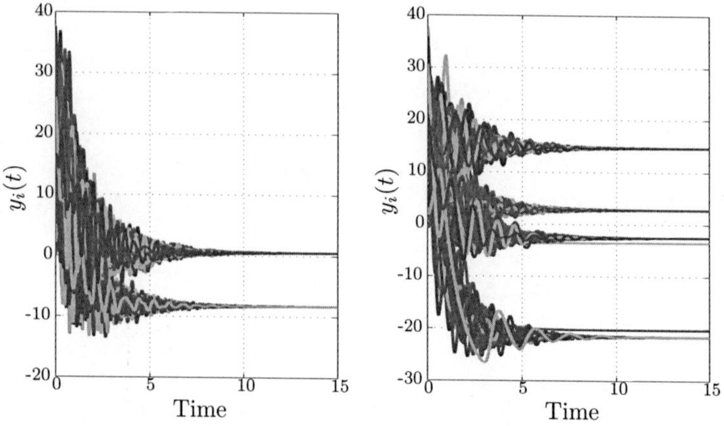

Figure 5.1: Preview of clustering or partial output agreement. Trajectories of two exemplary dynamical networks reaching agreement in two (left) or six (right) clusters.

and which connections are critical, in the sense that the network is likely to partition along them. Such an analysis is particularly difficult if the number of partitions is not specified. This motivates to search for *hierarchical clustering structures* in dynamical networks. We will see that this problem of community detection in dynamic networks (see e.g., Fortunato (2010)) can be tackled in the context of clustering as presented here.

The duality relations we derived in the previous chapter are the right framework to understand and to explain clustering in dynamical networks. We will identify in this chapter a class of dynamical networks, consisting of equilibrium independent dynamical systems, that exhibit a clustering behavior, and we will provide a precise analysis of the clustering structures. The novel duality theory will lead us the way to these results. The important connection we are going to exploit is the one between the optimal flow problem and the coupling nonlinearities of the dynamical network. When studying the static network flow framework, we will see that a clustering behavior appears in the potential variables of the network optimization problems if the flows are constrained. While the clustering structure of the optimal potential variables is not directly obvious in the original network optimization problems, it is easily revealed by a novel optimization problem that is in between the two dual problems, i.e., a saddle-point problem involving the potential and the flow variables. Now, as we can establish an explicit connection between the network optimization problems and the dynamical networks, it comes as no surprise that this clustering behavior is also reflected in the asymptotic behavior of certain dynamical networks. We consider therefore again networks of equilibrium independent passive systems with nonlinear, dynamic couplings. By exploiting the duality relations, we will see that the dynamic counterparts of the flow constraints are *saturated couplings.* Thus, networks of EIP systems with bounded interaction rules show asymptotically a clustering behavior.

We discuss the clustering phenomenon in this chapter from the two complementary perspectives: a static network optimization and a dynamic cooperative control perspective. In Section 5.2, we consider first the static network optimization framework and show that the addition of flow constraints leads to a clustering of the potential variables. We also characterize the situations in which a non-trivial clustering appears. While the optimization-based interpretation of clustering is interesting at its own right, it becomes particularly powerful as we can connect it to the passivity-based cooperative control framework. We establish this connection in Section 5.3, by exploiting the network theoretic interpretation of passivity-based control problems presented in the previous chapter. A Lyapunov analysis is presented to prove that the outputs of the dynamical network converge to the clustered potential variables, computed by the network optimization problems. Following this, it is shown how the proposed clustering analysis leads directly to a convex optimization framework for a novel hierarchical clustering algorithm, presented in Section 5.4. First, some combinatorial aspects of the appearing clustering structures are investigated before then a novel algorithm for identifying hierarchical clustering structures in dynamical networks is proposed.

5.2 Constrained Flows & Network Clustering

A major advantage of the network optimization framework derived in the previous chapter is that it allows us to integrate constraints. A typical limitation appearing in network flow problems are capacity constraints on the edges. As an ultimate goal, we aim to understand the dynamic counterpart of the constrained network flow problems. However, before we turn our attention to the dynamic system, we consider first the static versions of the network optimization problems with flow constraints and show that a combination of the two optimization problems (i.e., optimal flow and optimal potential problem) can provide important insights into the solution structure.

5.2.1 A Primal/Dual and Saddle-Point Perspective

We consider here the optimal flow problem (OFP1), but assume additionally that each edge is assigned a capacity of $w_k \in \mathbb{R}_{>0}$. That is, we restrict the flows to be contained in the set

$$\mathcal{W}_k = \{\mu_k : -w_k \leq \mu_k \leq w_k\}.$$

Optimal Flow Problem: Following the derivation of of the previous chapter we can directly formulate the network flow problem with additional flow constraints as

$$\begin{aligned} \min_{\mathbf{u},\mu} \quad & \sum_{i=1}^{n} K_i(\mathrm{u}_i) \\ & \mathbf{u} + E\mu = 0 \\ & -w_k \leq \mu_k \leq w. \end{aligned} \tag{OFP3}$$

To formulate (OFP3) in the standard form (4.5), we can add the indicator functions $\sum_{k=1}^{m} I_{\mathcal{W}_k}(\mu_k)$ for the constraint sets $\mathcal{W}_k$ in the cost to enforce the constraints. From this standard problem formulation, the dual problem can then be derived directly, taking into account that the convex conjugates of the indicator functions for the sets $\mathcal{W}_k$ are the weighted absolute values, i.e., $I^{\star}_{\mathcal{W}_k} = \sup\{\zeta_k \mu_k - I_{\mathcal{W}_k}(\mu_k)\} = w_k|\zeta_k|$.

Optimal Potential Problem: The dual to the optimal network flow problem (OFP3) is then the following *optimal potential problem*

$$\begin{aligned} \min_{\mathbf{y},\zeta} \quad & \sum_{i=1}^{n} K_i^{\star}(\mathrm{y}_i) + \sum_{k=1}^{m} w_k|\zeta_k| \\ & \zeta = E^{\top}\mathbf{y}. \end{aligned} \tag{OPP3}$$

Note that adding the capacity constraints allows the tensions ζ in (OPP3) to be unequal to zero.

Going directly to the dual using the convex conjugate representation seems fairly straight forward. However, performing this step directly conceals some important insights. In particular, the two optimization problems (OFP3), (OPP3) provide only little insights into the structure of the optimal solution. As we might believe after our previous discussion, that the potentials correspond to the asymptotic behavior of a cooperative dynamical network, we are interested in the structure of $\mathbf{y}$ as computed by (OPP3). Since the structure of $\mathbf{y}$ is not readily seen from (OPP3), we consider an additional representation of the network optimization problems. In fact, we consider an optimization problem which combines variables of the two primal/dual problem representations, i.e., on the one hand potential variables $\mathbf{y}$ and on the other hand flow variables μ.

Saddle-Point Problem: We consider now the following saddle-point problem

$$\max_{\mu_k} \min_{\mathrm{y}_i} \mathcal{L}(\mathbf{y},\mu) := \sum_{i=1}^{n} K_i^{\star}(\mathrm{y}_i) \; + \; \mu^{\top}E^{\top}\mathbf{y} \; - \; \sum_{k=1}^{m} I_{\mathcal{W}_k}(\mu_k). \tag{SPP}$$

The connection between the saddle-point problem and the original primal/dual pair of optimization problems follows from the next result.

Theorem 5.1. *Let* $(\mathbf{y},\mu)$ *be a saddle-point solution to* (SPP), *then* $\mathbf{y}$ *is an optimal solution to* (OPP3) *and* μ *is an optimal solution to* (OFP3).

Proof. From min-max theory follows directly that a saddle-point solution to (SPP) is such that $\mathbf{y}$ is a minimizer of the function $r(\mathbf{y}) = \max_{\mu} \mathcal{L}(\mathbf{y},\mu)$, and μ is a maximizer of the problem function $s(\mu) = \min_{\mathbf{y}} \mathcal{L}(\mathbf{y},\mu)$. First, note that we can introduce in (SPP) the variable $\zeta = E^{\top}\mathbf{y}$ and obtain $r(\mathbf{y}) = \sum_{i=1}^{n} K_i^{\star}(\mathrm{y}_i) + \max_{\mu_k} \sum_{k=1}^{m}\{\mu_k\zeta_k - I_{\mathcal{W}_k}(\mu_k)\} = \sum_{i=1}^{n} K_i^{\star}(\mathrm{y}_i) + \sum_{k=1}^{m} w_k|\zeta_k|$, which is exactly (OPP3). Alternatively, we could introduce in (SPP) the variable $\mathbf{u} = -E\mu$. We obtain then $s(\mu) = \sum_{i=1}^{n} \min_{\mathrm{y}_k}\{K_i^{\star}(\mathrm{y}_i) - \mathbf{u}^{\top}\mathbf{y}\} - \sum_{k=1}^{m} I_{\mathcal{W}_k}(\mu_k) = -\sum_{i=1}^{n} K_i(\mathrm{u}_i) - \sum_{k=1}^{m} I_{\mathcal{W}_k}(\mu_k)$. Since $\arg\max_{\mu} s(\mu) = \arg\min_{\mu} -s(\mu)$, the statement follows directly. □

The main advantage of considering the saddle-point problem (SPP) is that it gives us insights into the structure of the solutions. We will analyze next the solutions of the saddle-point problem and show that they exhibit in fact a clustering structure.

5.2.2 Saddle-Point Problem and Network Clustering

We consider now the saddle-point problem (SPP), which we rewrite for convenience as

$$\max_{\mu_k} \min_{y_i} \sum_{i=1}^{n} K_i^\star(y_i) \; + \; \mu^\top E^\top \mathbf{y}$$
$$\mu_k \in \mathcal{W}_k, \quad k \in \mathbf{E}.$$

In the following, we will sometimes use the short hand notation for the constraint set $\mathcal{W} = \mathcal{W}_1 \times \cdots \times \mathcal{W}_m$. In order to get a first intuition about the solution structure of (SPP), we can consult the game-theoretic interpretation of the problem. Assume there are two different types of decision makers. The first class of decision makers is placed on the nodes of $\mathcal{G}$ and aims to minimize the individual cost functions $K_i^\star(y_i)$ by choosing the variables y_i accordingly. The second class of decision makers is placed on the edges of $\mathcal{G}$. A decision maker in this class penalizes any deviation between the decision variables of its incident nodes. That is, the edge decision makers benefit from any disagreement between the node decision makers, and they adjust their decision variables μ_k accordingly. Following this interpretation, it is obvious that the network nodes will be forced to agreement, if no constraints are imposed on the edge variables μ_k. In fact, this corresponds to the situation without flow constraints, discussed in Chapter 4. We know from the discussion in Chapter 4 that the network without flow constraints will, in fact, reach complete agreement on the potential variables. However, the saddle-point problem formulation (SPP) does not permit the edge decision makers to arbitrarily penalize the deviation between neighboring agents, since the penalty variables μ_k are restricted to be contained in the set $\mathcal{W}_k$. This additional constraint has a strong impact on the structure of the primal solution $\mathbf{y}$. Next, we investigate this impact on the solution structure.

Saddle Points In a first step, we formally introduce the notion of a saddle-point and discuss some properties.

Definition 5.2. *A point* $(\bar{\mathbf{y}}, \bar{\mu})$ *is a* saddle-point *of* (SPP) *if*

$$\mathcal{L}(\bar{\mathbf{y}}, \mu) \leq \mathcal{L}(\bar{\mathbf{y}}, \bar{\mu}) \leq \mathcal{L}(\mathbf{y}, \bar{\mu}), \quad \forall\, \mathbf{y} \in \mathbb{R}^n,\; \mu \in \mathcal{W}_1 \times \cdots \times \mathcal{W}_m.$$

We will denote the *set of all saddle-points* to (SPP) in the following by $\mathcal{S}^* = \mathcal{Y}^* \times \mathcal{M}^*$. Since the constraint set $\mathcal{W}$ is nonempty, convex, and compact, and the function $\mathcal{L}(\mathbf{y}, \mu)$ is convex for each fixed $\mu \in \mathcal{W}$, and concave for each $\mathbf{y} \in \mathbb{R}^n$ with bounded level sets, the set of saddle-points $\mathcal{S}^*$ for (SPP) is non-empty, see e.g., (Bertsekas, 2009, Proposition 4.7). We can come directly to the next observation, which is that there is a unique vector $\bar{\mathbf{y}}$ at which a saddle-point can be attained, whereas the component μ is not uniquely defined. In fact, note that, by assumption, the functions $K_i^\star(y_i)$ are strictly convex, leading to the uniqueness result for the $\mathbf{y}$ component. However, it can be easily seen that any feasible μ can be varied in $\mathcal{N}(E)$. Therefore, let $(\bar{\mathbf{y}}, \bar{\mu})$ be a saddle-point of (SPP), then

$$\mathcal{Y}^* = \{\bar{\mathbf{y}}\} \quad \text{and} \quad \mathcal{M}^* = \{\mu \in \Gamma \,|\, \mu = \bar{\mu} + \nu,\; \nu \in \mathcal{N}(E)\}.$$

Recall that $\mathcal{N}(E)$ is the circulation or flow space of $\mathcal{G}$ and is defined by the cycles of the graph. Thus, the set $\mathcal{M}^*$ contains more than one point if and only if $\mathcal{G}$ contains at least one cycle. This observation matches clearly the interpretation of μ as a flow vector. Given a fixed divergence, the flows can only be varied along the cycles of the network. If $\mathcal{G}$ is a tree, the flows are uniquely defined.

For completeness, we review also the first-order optimality conditions for the saddle-point problem. Let $(\bar{\mathbf{y}}, \bar{\mu})$ be a saddle-point, then

$$\nabla \mathbf{K}^{\star}(\bar{\mathbf{y}}) + E\bar{\mu} = 0, \quad \text{and} \quad \bar{\mathbf{y}}^{\top} E(\mu - \bar{\mu}) \leq 0, \quad \forall\, \mu \in \mathcal{W}. \tag{5.1}$$

Network Clustering Having established the existence and uniqueness properties of the saddle-points for (SPP), we now show how these solutions lead to clusters in the graph $\mathcal{G}$. First, we introduce the notion of a *saturated edge* in the graph.

Definition 5.3. *An edge $k \in \mathbf{E}$ is said to be* saturated *if for all $\bar{\mu} \in \mathcal{M}^*$ it holds that $\bar{\mu}_k \in \mathrm{bd}\mathcal{W}_k$.*

We use here the notation $\mathrm{bd}\mathcal{W}_k$ for the boundary of the set $\mathcal{W}_k$. Similarly, we will write $\mathrm{int}\mathcal{W}_k$ for the interior of $\mathcal{W}_k$. The wording of a "saturated edge" follows directly from the interpretation of μ as a flow. Note that in general $\bar{\mu}_k \in \mathrm{bd}\mathcal{W}_k$ for a particular $\bar{\mu}$ does not imply the edge is saturated. For an edge to be saturated, the constraint associated with that edge must be active for *all* possible saddle-points in the set $\mathcal{M}^*$. That is, for any feasible flow through the network, there is an edge k that is at its capacity bounds. The following lemma connects the definition of saturated edges to graph properties.

Lemma 5.4. *Any cycle in $\mathcal{G}$ contains either none or at least two saturated edges.*

Proof. Assume by contradiction that edge $k \in \mathbf{E}$ is the only saturated edge contained in a cycle with a corresponding signed path vector $\mathbf{s}$. Then $\mathrm{s}_k \neq 0$ and from Theorem 4.3, $\mathbf{s} \in \mathcal{N}(E)$. From the structure of the set $\mathcal{M}^*$ follows that there exists a $\delta \in \mathbb{R}$ sufficiently small such that $\tilde{\mu} = \bar{\mu} + \delta\mathbf{s} \in \mathcal{M}^*$ and $\tilde{\mu}_k \in \mathrm{int}\mathcal{W}_k$. But this contradicts the definition of a saturated edge. Therefore, k cannot be saturated. Thus, if a cycle contains a saturated edge, it must contain at least two saturated edges. □

In fact, if the set $\mathcal{M}^*$ contains saturated edges, then there is a corresponding cut-set for the graph $\mathcal{G}$ comprised of those edges.

Lemma 5.5. *The set of saturated edges in $\mathcal{M}^*$ forms a cut-set for the graph.*

Proof. First, assume that an edge k in the graph is saturated and is *not* contained in any cycle in $\mathcal{G}$. Then $\bar{\mu}_k \in \mathrm{bd}\mathcal{W}_k$, and its deletion from the graph must result in an increase in the number of components, thus forming a cut-set. Now assume that a saturated edge k is contained in one or more cycles. Then by Lemma 5.4 any cycle contains at least one other saturated edge. The deletion of two or more edges from a cycle results in an increase in the number of components in the graph, and thus each saturated edge in a cycle is included in a cut-set. □

Lemma 5.5 makes a strong connection between the saddle-points of (SPP), saturated edges, and cut-sets. The equivalent interpretation in the context of flow networks is as follows. If the maximum of flow is transported through a network, then there must be a set of edges, forming a cut-set, such that the flow on these edges is on the capacity limits. We are now able to state the main result of this section, relating clustering to saddle-points.[1]

Theorem 5.6 (Saddle-Point Clustering). *Let $\mathcal{S}^* = \{\bar{\mathbf{y}}\} \times \mathcal{M}^*$ be the saddle-points of (SPP), and let $\mathbf{Q} \subseteq \mathbf{E}$ be the set of saturated edges. Then $\mathbf{Q}$ induces a p-partition $\mathbb{P} = \{\mathbf{P}_1, \ldots, \mathbf{P}_p\}$ and each cluster $\mathcal{P}_i$ induced by the set $\mathbf{P}_i$ is in exact agreement.*

Proof. Let $(\bar{\mathbf{y}}, \bar{\boldsymbol{\mu}}) \in \mathcal{S}^*$ be a saddle-point with $\bar{\mu}_k \in \text{int}\mathcal{W}_k$ for all non-saturated edges. Note that stating cluster $\mathcal{P}_l$ is in agreement is equivalent to $E(\mathcal{P}_l)^\top \bar{\mathbf{y}}(\mathcal{P}_l) = 0$. Assume, in order to arrive at a contradiction, that there exists some $P_l \in \mathbb{P}$ such that $E(\mathcal{P}_l)^\top \bar{\mathbf{y}}(\mathcal{P}_l) \neq 0$. Denote by $\mathcal{Q}$ the subgraph induced by $\mathbf{Q}$. The function (SPP) can then be written as

$$\mathcal{L}(\bar{\mathbf{y}}, \bar{\boldsymbol{\mu}}) = \sum_{i=1}^{n} K_i^\star(\bar{y}_i) + \sum_{l=1}^{p} \bar{\boldsymbol{\mu}}(\mathcal{P}_l)^\top E(\mathcal{P}_l)^\top \bar{\mathbf{y}}(\mathcal{P}_l) + \bar{\boldsymbol{\mu}}(\mathcal{Q})^\top E(\mathcal{Q})^\top \bar{\mathbf{y}}(\mathcal{Q}). \tag{5.2}$$

Since all clusters except $\mathcal{P}_i$ are assumed to be in agreement, the second summand of (5.2) can be written as $\sum_{k=(i,j)\in\mathcal{P}_l} \bar{\mu}_k(\bar{y}_i - \bar{y}_j)$. Assume without loss of generality that only the edge $k = (i,j)$ in $\mathcal{P}_l$ connects two nodes, that are not in agreement with a positive difference (e.g., $\bar{y}_i - \bar{y}_j > 0$). Then there exists an $\epsilon > 0$ such that $\bar{\mu}_k + \epsilon \in \mathcal{W}_k$ and $\bar{\mu}_k(\bar{y}_i - \bar{y}_j) < (\bar{\mu}_k + \epsilon)(\bar{y}_i - \bar{y}_j)$. Let $\tilde{\boldsymbol{\mu}}$ be the edge value after adding ϵ to only one $\bar{\mu}_k$ as described above. Then $\mathcal{L}(\bar{\mathbf{y}}, \bar{\boldsymbol{\mu}}) \leq \mathcal{L}(\bar{\mathbf{y}}, \tilde{\boldsymbol{\mu}})$. This contradicts the assumption that $(\bar{\mathbf{y}}, \bar{\boldsymbol{\mu}})$ is a saddle-point. Therefore, each cluster $\mathcal{P}_i$ must be in agreement. □

With this theorem, we can directly make a conclusion on the clustering structure of the network. The network will partition along the saturated edges contained in the saddle-points of (SPP). The clusters $\mathcal{P}_i$ are the connected components of the graph $\mathcal{G}$ after deleting all the saturated edges.

Let us briefly think about this result in the context of general network theory. The last result can be interpreted as follows. Consider the optimal flow problem (OFP3). This problem aims to optimize the in/outflow to the network according to the cost functions $K_i(u_i)$. If the optimal in/outflow of the network can be realized without violating the capacity constraint on the transportation lines, then all nodes in the network will attain the same potential. However, if the optimal in/out-flow is limited by the capacity of the transportation networks, then the potential variables will form clusters, in the sense that all nodes connected by a line not at the capacity limit have the same potential, while the potentials of nodes connected by saturated lines are different. We want to emphasize that the clustering of the potential variables could be seen fairly directly from the saddle-point problem, while it was not directly obvious from the original problem formulations.

5.3 Clustering in Dynamical Networks

While the clustering behavior of the potential variables in constrained network flow problems is interesting on its own right (we will come back to this in the next section), our main

[1] For clarity of the presentation, we will in the following proof use a slightly different notation. We will write for the incidence matrix of graph $\mathcal{G}$ the longer notation $E(\mathcal{G})$ instead of simply E. This helps to distinguish the incidence matrices associated to different graphs. Furthermore, for vectors defined in the node or edge space of the graph, we will denote those components corresponding to a subgraph, e.g., $\mathcal{G}' \subset \mathcal{G}$, with $\mathbf{y}(\mathcal{G}')$.

interest in this chapter is the clustering of dynamical networks. Motivated by the clustering trajectories shown in Figure 5.1, we want to characterize in this section a class of dynamical networks that exhibit such a behavior. Therefore, we will introduce the dynamic counterpart of the constrained network flow problems, following similar lines of argumentation as in Chapter 4.

5.3.1 A Dynamical Model for Clustering

We start again with the same passivity-based cooperative control framework, as introduced in Chapter 4. That is, we consider again a network of dynamical systems, with each node representing an output strictly equilibrium independent passive SISO system

$$\begin{aligned} \dot{x}_i(t) &= f_i(x_i(t), u_i(t)) \\ y_i(t) &= h_i(x_i(t), u_i(t)), \end{aligned} \qquad i \in \mathbf{V}. \tag{5.3}$$

The nodes are coupled by the same integral control structure, i.e.,

$$\begin{aligned} \dot{\boldsymbol{\eta}}(t) &= E^\top \boldsymbol{y}(t), \quad \boldsymbol{\eta}(t_0) \in \mathcal{R}(E^\top), \\ \boldsymbol{u}(t) &= -E\boldsymbol{\psi}(\boldsymbol{\eta}(t)). \end{aligned} \tag{5.4}$$

For the following discussion, we impose the same assumptions as in Chapter 4, and we do not review these assumptions here. However, we want to recall that the coupling nonlinearities ψ_k are assumed to be the gradients of some convex functions P_k. These coupling nonlinearities are the key elements that will cause the clustering behavior. In fact, the dynamical model considered in this section will differ from the model studied in Chapter 4 only by the choice of the coupling functions.

We want to define now a version of the cooperative control framework (5.3), (5.4) that corresponds to the constrained network flow problems (OFP3), (OPP3). Since we have already seen that the constrained network flow problems exhibit a clustering behavior, we can expect a similar behavior for their dynamic counterparts. In order to define the dynamic counterpart, have to define the coupling nonlinearities in (5.4) in a suitable way. We can remember now that the connection between the optimal flows and the coupling controller was fully found in the optimization problem (OFP2). That is, the dynamic variables $\boldsymbol{\mu}(t)$ approach asymptotically the solution to the optimal flow problem (OFP2), and the cost function of (OFP2) were the convex conjugates $P_k^\star$ of the integral functions of the coupling nonlinearities.

We can start with this observation for the controller design. It is a classical trick in optimization theory, to extend the objective function such that it takes value infinity outside the feasible set. For incorporating the flow constraints into (OFP2), we can consider cost functions, which are only finite for feasible flows and are infinity for infinite flows,[2] i.e.,

$$P_k^\star(\mu_k) \begin{cases} < +\infty & \text{if } \mu_k \in \mathcal{W}_k \\ = +\infty & \text{if } \mu_k \notin \mathcal{W}_k. \end{cases}$$

However, as we are interested in the connection to dynamical systems, it is advantageous to impose some further smoothness assumptions. In this direction, we consider now a new class of flow cost functions.

[2]That is, we consider cost functions with effective domain corresponding to the capacity of an edge.

Assumption 5.1. *The flow cost functions $P_k^\star(\mu_k)$ are essentially smooth, even, strictly convex functions on the domain $\mathcal{W}_k = \{\mu_k : -w_k \leq \mu_k \leq w_k\}$ for some edge capacity $w_k > 0$, i.e.,*

$$\lim_{|\mu_k| \to w_k} P_k^\star(\mu_k) \to \infty.$$

For a definition of essentially smooth functions we refer to Appendix A. These flow cost functions will lead the way to the desired coupling nonlinearities. The flow cost functions are the convex conjugates of the integral functions P_k of the coupling nonlinearities, i.e., $\psi_k(\eta_k) = \nabla P_k(\eta_k)$.

Lemma 5.7. *Let Assumption 5.1 hold and let P_k be the convex conjugate of $P_k^\star$. Then the functions $\psi_k(\cdot) = \nabla P_k(\cdot)$ are odd, monotone and* bounded, *i.e.,*

$$\lim_{\|s\| \to \infty} \|\psi_k(s)\| \to w_k.$$

Proof. By assumption P_k is essentially smooth. If $P_k^\star$ is convex the same holds for P_k, and thus ψ_k is monotone. By assumption $P_k^\star$ is odd. Thus $P_k(-\eta_k) = \sup_{\mu_k}\{-\eta_k \mu_k - P_k^\star(\mu_k)\} = \sup_{\mu_k}\{\eta_k(-\mu_k) - P_k^\star(-\mu_k)\} = \sup_{\tilde{\mu}_k}\{\eta_k \tilde{\mu}_k) - P_k^\star(\tilde{\mu}_k)\} = P_k(\eta_k)$. Thus, $P_k(\eta_k)$ is odd and, consequently $\psi_k(\eta_k)$ is even. To see boundedness of the controller output trajectories, note that $\lim_{|\mu_k| \to w_k} P_k^\star(\mu_k) \to \infty$ implies, by differentiability, that $\lim_{|\mu_k| \to w_k} |\nabla P_k^\star(\mu_k)| \to \infty$. Clearly, the domain of $\nabla P_k^\star(\mu_k)$ is $\mathcal{W}_k$ and the image is $\mathbb{R}$. By assumption $\nabla P_k^\star(\mu_k)$ is invertible, and we can define $\eta_k = \nabla P_k^\star(\mu_k) \Leftrightarrow \nabla P_k(\eta_k) = \mu_k$, since $(\nabla P_k^\star)^{-1} = \nabla P_k$. The domain of ∇P_k is $\mathbb{R}$, while its image is $\mathcal{W}_k$. It follows now directly that, whenever $|\nabla P_k(\eta_k)| \to w_k$, the corresponding variable must satisfy $|\eta_k| \to \infty$. By monotonicity and continuity follows also that $\nabla P_k^\star(\mu_k) < \infty$ for any $\mu_k \in \mathrm{int}\mathcal{W}_k$. Thus, we can directly conclude $\lim_{|\eta_k| \to \infty} |\nabla P_k(\eta_k)| \to w_k$. □

Example 5.8. *One example of such a C^2 function, being finite only within the capacity constraints, is*

$$P_k^\star(\mu_k) = \frac{1}{2} w_k \log(1 - \mu_k^2) + w_k \mu_k \tanh^{-1}(\mu_k). \tag{5.5}$$

This is the integral function of $w_k \tanh^{-1}(\mu_k)$. Consequently, we have that

$$\psi_k(\eta_k(t)) = \nabla P_k(\eta_k(t)) = w_k \tanh(\eta_k(t)),$$

i.e., a monotonic and bounded function. To complete the picture, we want to point out that the integral of the coupling function is

$$P_k(\eta_k) = \ln \cosh(\eta_k).$$

The flow cost function $P_k^\star$, the coupling nonlinearity $\psi_k(\eta)$ and the integral function $P_k(\eta_k)$ are illustrated in Figure 5.2.

In fact, under the new assumption, the coupling nonlinearities will be saturated functions, such as e.g., $\psi_k(\eta_k) = w_k \tanh(\eta_k)$. Note that the new Assumption 5.1 replaces our previous Assumption 4.2, where we required the functions $P_k(\eta_k)$ to be strongly convex. In fact, the functions $P_k(\eta_k)$ now have the property that they asymptotically approach an affine function. Thus, they are convex but no longer strongly convex.

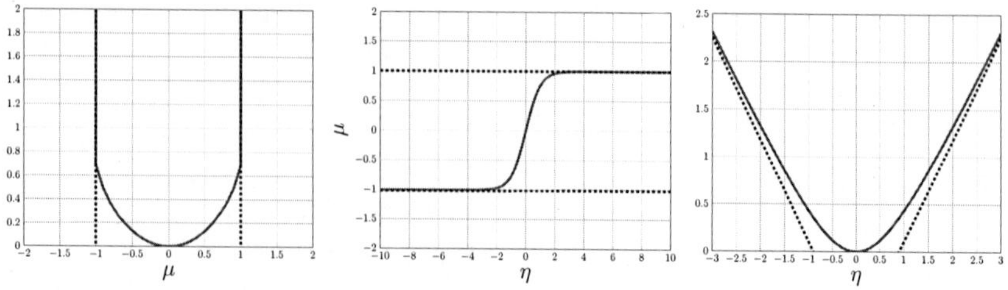

Figure 5.2: Relation between the *flow cost function* $P^\star(\mu)$, the *coupling nonlinearity*, here $\psi(\eta) = \nabla P(\eta) := \tanh(\eta)$, and the *coupling function integral* $P(\eta)$. Note that the image of the coupling nonlinearity $\psi(\eta)$, i.e., $(-1, 1)$, corresponds to the effective domain of $P^\star(\mu)$.

5.3.2 Clustering Analysis and Convergence

The dynamical network counterpart to the constrained network flow problems (OFP3), (OPP3) is the dynamical network (4.16), (4.24) with *bounded coupling functions.* Now, in the same way as the flow constraints caused a clustering of the node potentials in the static network optimization framework, the bounded coupling functions cause a *clustering,* or *partial output agreement,* of the dynamical network.

Definition 5.9. *A network of dynamical systems is said to be* clustering, *or reaching partial output agreement, if there is a partition of the node set* $\mathbb{P} = \{\mathbf{P}_1, \ldots, \mathbf{P}_p\}$, $(1 \leq p \leq n)$, *s.t.*

$$\lim_{t\to\infty} y_i(t) \to \beta_l \ \text{ for all } i \in \mathbf{P}_l$$

for some $\beta_l \in \mathbb{R}$ *with* $\beta_l \neq \beta_q$ *for* $l \neq q$.

The asymptotic behavior of the dynamical network (4.16), (4.24) with *bounded* coupling functions can now be connected to the network flow problems (OFP3), (OPP3). We establish this connection along the same line of argumentation as in the proof of Theorem (4.21), but need some additional technical details. We first establish the existence of a steady state solution.

Lemma 5.10. *Let Assumptions 4.1 and 5.1 hold and let* $(\mathbf{u}, \mu)$ *and* $(\mathbf{y}, \zeta)$ *be solutions to* (OFP3) *and* (OPP3), *respectively. Then,* $\mathbf{u}$ *and* $\mathbf{y}$ *are an equilibrium input-output pair for the systems* (5.3), *i.e.,* $\mathbf{k}_\mathrm{y}(\mathbf{u}) = \mathbf{y}$.

The statement follows directly from the duality relation between (OFP3) and (OPP3), and the definition of the objective functions $K_i(\mathrm{u}_i)$. Recall, that the inputs of the dynamical network are given by $\boldsymbol{u}(t) = -E\psi(\boldsymbol{\eta}(t))$ with $\boldsymbol{\eta}(t) \in \mathcal{R}(E^\top)$. Now, if the coupling nonlinearities approach a limit, it might not be possible to find a vector $\eta \in \mathcal{R}(E^\top)$ with only finite entries that realizes the desired equilibrium input $\mathbf{u} = -E\psi(\eta)$. However, we show next that one can always find a trajectory such that the equilibrium input is approached asymptotically. In fact, the unboundedness of η corresponds to the clustering of the dynamical network, since we have in the dynamic counterpart $\dot{\boldsymbol{\eta}}(t) = E^\top \boldsymbol{y}(t)$, while in the steady state $E^\top \mathbf{y} \neq 0$.

Lemma 5.11. *Let* $(\mathbf{u}, \boldsymbol{\mu})$ *and* $(\mathbf{y}, \boldsymbol{\zeta})$ *be solutions to* (OFP3) *and* (OPP3), *respectively. Then there exists a sequence* $\{\boldsymbol{\eta}^\ell\}$ *with* $\boldsymbol{\eta}^\ell \in \mathcal{R}(E^\top)$ *and* $\ell \in \mathbb{Z}_+$ *such that*

$$\lim_{\ell\to\infty} \left(\mathbf{u} + E\nabla\psi(\boldsymbol{\eta}^\ell)\right) \to 0.$$

The proof of this result relies on the optimal potential problem (OPP2), which we previously used to determine the steady state of the controller variable.

Proof. Consider the problem (OPP2) for some $\mathbf{u} \in \mathbb{R}^n$,

$$\min_{\boldsymbol{\eta},\mathbf{v}} \quad \sum_{k=1}^{m} P_k(\eta_k) + \sum_{i=1}^{n} \mathrm{u}_i \mathrm{v}_i, \quad \text{s.t.} \quad \boldsymbol{\eta} = E^\top \mathbf{v}.$$

Replacing the divergence $\mathbf{u}$ with the corresponding flow representation $\mathbf{u} = -E\boldsymbol{\mu}$ and inserting the constraint in the objective function leads to the alternative representation of the problem as

$$\min_{\boldsymbol{\eta}} \quad \sum_{k=1}^{m} P_k(\eta_k) - \boldsymbol{\mu}^\top \boldsymbol{\eta}, \quad \boldsymbol{\eta} \in \mathcal{R}(E^\top). \tag{5.6}$$

Note that (5.6) has a finite optimal solution whenever $\boldsymbol{\mu}$ strictly satisfies all capacity constraints, that is $\boldsymbol{\mu} \in \operatorname{int}\left\{\mathcal{W}_1 \times \cdots \times \mathcal{W}_m\right\}$.

Now, define a sequence $\{\boldsymbol{\mu}^\ell\}$ such that $\lim_{\ell\to\infty} \boldsymbol{\mu}^\ell \to \boldsymbol{\mu}$, while any vector $\boldsymbol{\mu}^\ell$ is such that each entry μ_k^ℓ strictly satisfies the constraints, i.e., $-w_k < \mu_k^\ell < w_k$. Each problem has a finite optimal solution since $\boldsymbol{\mu}^\ell$ is strictly in the range of the coupling function $\nabla \mathbf{P}(\boldsymbol{\eta}) = \psi(\boldsymbol{\eta})$. Mimicking the proof of Theorem 4.19, we can show that for each $\boldsymbol{\mu}^\ell$, the optimal solution to (5.6) satisfies $\mathbf{u}^\ell = -E\psi(\boldsymbol{\eta}^\ell)$, where $\mathbf{u}^\ell = -E\boldsymbol{\mu}^\ell$. From $\lim_{\ell\to\infty} \boldsymbol{\mu}^\ell \to \boldsymbol{\mu}$ follows $\lim_{\ell\to\infty} \mathbf{u}^\ell \to \mathbf{u}$, and consequently $\lim_{\ell\to\infty}(\mathbf{u} + E\nabla\psi(\boldsymbol{\eta}^\ell)) \to 0$. □

We are now ready to connect the asymptotic behavior of the dynamical network (5.3), (5.4) with bounded coupling functions to the constrained network flow problems (OFP3), (OPP3). A natural approach to analyze these dynamics is to consider again the Lyapunov analysis of Section 4.3.3. An important component of the Lyapunov analysis was the use of the Bregman distance $B_P(\boldsymbol{\eta}(t), \boldsymbol{\eta})$ as defined in (4.28). Unfortunately, as we discussed in Lemma 5.11, there is not necessarily a finite steady state for the internal model controller, and consequently $\boldsymbol{\eta}$ might have unbounded components. Thus, we cannot simply consider $\mathbf{B}_P(\boldsymbol{\eta}(t), \boldsymbol{\eta})$ as storage function. However, we can now use the sequence $\{\boldsymbol{\eta}^\ell\}$, which we constructed in the proof of Lemma 5.11, and consider

$$\bar{\mathbf{B}}_P(\boldsymbol{\eta}(t)) := \lim_{\ell\to\infty} \mathbf{B}_P(\boldsymbol{\eta}(t), \boldsymbol{\eta}^\ell) = \lim_{\ell\to\infty} \mathbf{P}(\boldsymbol{\eta}(t)) - \mathbf{P}(\boldsymbol{\eta}^\ell) - \nabla\mathbf{P}(\boldsymbol{\eta}^\ell)^\top(\boldsymbol{\eta}(t) - \boldsymbol{\eta}^\ell). \tag{5.7}$$

The first observation we make follows directly from Lemma 5.11, namely $\lim_{\ell\to\infty} \nabla\mathbf{P}(\boldsymbol{\eta}_\ell) = \boldsymbol{\mu}$. Next, we exploit the property that the functions $P_k(\eta_k)$ have an affine function as an asymptote, which is a consequence of Assumption (5.1). The symmetry and the existence of an asymptote implies that $\lim_{|\eta_k|\to\infty} P_k(\eta_k) + \nabla P_k(\eta_k)^\top \eta_k = \text{const}$. Since we either have $\lim_{\ell\to\infty} \eta_k^l$ approaches a finite value or diverges, we can conclude that

$$\lim_{\ell\to\infty} \mathbf{P}(\boldsymbol{\eta}^\ell) - \nabla\mathbf{P}(\boldsymbol{\eta}^\ell)^\top \boldsymbol{\eta}^\ell = c \tag{5.8}$$

for some constant c. From these observations, we can conclude that $\bar{\mathbf{B}}_P(\boldsymbol{\eta}(t)) := \mathbf{P}(\boldsymbol{\eta}(t)) - \boldsymbol{\mu}^\top \boldsymbol{\eta}(t) - c$ is a non-negative function, with minimal value zero, which is attained as $\boldsymbol{\eta}(t) \to \lim_{\ell\to\infty} \boldsymbol{\eta}^\ell$. Using the same argumentation steps as in (4.26), we obtain that

$$\dot{\bar{\mathbf{B}}}(\boldsymbol{\eta}(t)) = -(\boldsymbol{y}(t) - \mathbf{y})^\top(\boldsymbol{u}(t) - \mathbf{u}) - \mathbf{y}^\top(\boldsymbol{u}(t) - \mathbf{u}). \tag{5.9}$$

In contrast to the discussion of Section 4.3.3, we have now that in general $\mathbf{y} \notin \operatorname{span}\{\mathbf{1}\}$, as the potentials might have a clustered structure. However, exploiting the duality between the potential and divergence variables, we can equivalently write $\mathbf{y}^\top(\boldsymbol{u}(t)-\mathbf{u}) = \nabla\mathbf{K}(\mathbf{u})^\top(\boldsymbol{u}(t)-\mathbf{u})$. Here, $\boldsymbol{u}(t)$ is any feasible solution to (OFP3), and $\mathbf{u}$ is an optimal solution. Thus, an optimality condition of (OFP3) is that

$$\nabla\mathbf{K}(\mathbf{u})^\top(\boldsymbol{u}(t) - \mathbf{u}) \geq 0. \tag{5.10}$$

We are now ready to prove the convergence of the dynamical system.

Theorem 5.12. *Let Assumptions 4.1 and 5.1 hold and let* $(\mathbf{u}, \boldsymbol{\mu})$ *and* $(\mathbf{y}, \boldsymbol{\zeta})$ *be solutions to* (OFP3) *and* (OPP3), *respectively. Then,* $\lim_{t\to\infty} \boldsymbol{y}(t) \to \mathbf{y}$ *and* $\lim_{t\to\infty}(-E\nabla\mathbf{P}(\boldsymbol{\eta}(t))) \to \mathbf{u}$.

Proof. Consider the nonnegative function $W(\boldsymbol{x}(t), \boldsymbol{\eta}(t)) = \mathbf{S}(\boldsymbol{x}(t)) + \bar{\mathbf{B}}_P(\boldsymbol{\eta}(t))$ with $\mathbf{S}(\boldsymbol{x}(t)) := \sum_{i=1}^n S_i(x_i(t))$ the storage functions of the systems with respect to the equilibrium inputs $\mathbf{u}$. Using (4.9), (5.9), and (5.10), we directly obtain

$$\dot{W}(\boldsymbol{x}(t), \boldsymbol{\eta}(t)) \leq -\left(\sum_{i=1}^n \rho_i(y_i - \mathrm{y}_i)\right). \tag{5.11}$$

Integrating both sides and applying Barbalat's Lemma (Khalil, 2002), we finally conclude $\lim_{t\to\infty} \boldsymbol{y}(t) \to \mathbf{y}$. This can only hold if $\lim_{t\to\infty} \boldsymbol{u}(t) \to \mathbf{u}$, and thus $\lim_{t\to\infty}\big(E\nabla\mathbf{P}(\boldsymbol{\eta}(t)) + \mathbf{u})\big) \to 0$. □

Summarizing, the clustering of diffusively coupled networks of passive dynamical systems, caused by bounded coupling nonlinearities, can be understood by studying a constrained network flow problem (OFP3). In fact, the asymptotic clustering of the dynamical systems' outputs is equivalent to the clustering of the node potentials in the problem (OPP3).

5.3.3 Application Examples

The presented results have several practical implications. In particular, they provide a framework to efficiently analyze cooperative dynamical networks, since network flow optimization problems can be solved using numerical methods. We discuss the relevance of our results for different dynamical models in the following.

Analysis of a Traffic Dynamics Model: We discuss here how our methodology can be used for the analysis of a traffic dynamics model. We consider in the following a microscopic traffic model, and focus on the *optimal velocity model* proposed in Bando et al. (1995), Helbing and Tilch (1998). We consider some modifications of the original model: (i) the drivers are heterogeneous and have different "preferred" velocities, (ii) the influence between cars is bi-directional, and (iii) vehicles can overtake other vehicles directly. The model is as follows. Each vehicle adjusts its velocity v_i according to the control

$$\dot{v}_i = \kappa_i[V_i(\Delta\mathbf{p}) - v_i], \tag{5.12}$$

where the adjustment $V_i(\Delta\mathbf{p})$ depends on the relative position to other vehicles, i.e., $\Delta\mathbf{p}$, as

$$V_i(\Delta\mathbf{p}) = V_i^0 + V_i^1 \sum_{j\in\mathcal{N}(i)} \tanh(p_j - p_i). \tag{5.13}$$

Here p_i denotes the position of vehicle i and $\mathcal{N}(i)$ is used to denote the neighboring vehicles influencing vehicle i. Throughout this example, we assume that the set of neighbors to a vehicle is fixed according to the initial traffic configuration and does not change over time. The constants V_i^0 are "preferred velocities" and V_i^1 are sensitivities of the drivers. In the following we assume $V_i^0, V_i^1 > 0$ and $V_i^0 \neq V_j^0$ (heterogeneity).

We can now provide an interpretation of this traffic model in the context of network theory. The node dynamics can be directly identified as $\dot{v}_i(t) = \kappa_i[-v_i(t) + V_i^0 + V_i^1 u_i(t)]$, with $v_i(t)$ as node state and input $u_i(t) := \sum_{j\in\mathcal{N}(i)} \tanh(p_j(t) - p_i(t))$. Choosing simply the output $y_i(t) = v_i(t)$, we obtain an OSEIP system with the affine equilibrium input-output map $k_{\mathrm{y},i}(\mathrm{u}_i) = V_i^0 + V_i^1\mathrm{u}_i$. The *potential* variables are therefore the velocities of the vehicles, i.e., v_i, while the *divergence* at a node, i.e., vehicle, is the influence of the other vehicles, i.e., $\sum_{j\in\mathcal{N}(i)} \tanh(p_j - p_i)$. The objective functions, which determine the equilibrium, are here the quadratic functions

$$K_i(\mathrm{u}_i) = \frac{V_i^1}{2}\mathrm{u}_i^2 + V_i^0\mathrm{u}_i \quad \text{and} \quad K_i^\star(\mathrm{y}_i) = \frac{1}{2V_i^1}(\mathrm{y}_i - V_i^0)^2. \tag{5.14}$$

The state of the coupling controller can be readily identified as the relative position of the vehicles, i.e., $\boldsymbol{\eta}(t) := \Delta\mathbf{p}(t)$, while the controller output is $\mu_k(t) = \tanh(\eta_k(t)) = \tanh(p_j(t) - p_i(t))$. This gives us also the interpretation of the *tensions* as the relative positions of the vehicles and the *flows* as their mapping through the coupling functions. The problem corresponds to a network flow problem with unity capacity constraints on the edges, i.e., $-1 \leq \mu_k \leq 1$. With this model, the problem (OPP3) takes the form

$$\begin{aligned} \min_{\mathbf{y},\boldsymbol{\zeta}} \sum_{i=1}^{n} \frac{1}{2V_i^1}(\mathrm{y}_i - V_i^0)^2 &+ \sum_{k=1}^{m} |\zeta_k| \\ \boldsymbol{\zeta} = E^\top \mathbf{y}. \end{aligned} \tag{5.15}$$

Note that this problem has a very characteristic structure, i.e., it is a quadratic problem plus an additional ℓ_1 term penalizing the tensions $\boldsymbol{\zeta}$. Remarkably, convex optimization problems of this structure can be solved very efficiently using modern optimization algorithms.

The coupling functions satisfy Assumption 5.1. Therefore, Theorem 5.6 holds, and we can use the static network optimization problems to analyze the asymptotic behavior of the system. In particular, the potential problem (5.15) has a nice convex structure and can be solved efficiently by a variety of numerical algorithms.

Since the coupling functions are saturated, the model will show for certain configurations (i.e., for certain choices of V_i^0, V_i^1) a clustering behavior. Here, clustering means that the vehicles will form groups that travel with the same velocity, while the different groups have different velocities. Such a behavior can clearly be expected in a traffic system. A computational study with 100 vehicles placed on a line graph is shown in Figure 5.3. The sensitivity parameter is $\kappa = 0.6$ for all vehicles, while the parameters V_i^0 and V_i^1 are chosen as a common nominal parameter plus a random component, i.e., $V_i^0 = V_{nom}^0 + V_{i,rand}^0$. The common off-set is $V_{nom}^0 = 25\frac{m}{s}$ and $V_1 = 10\frac{m}{s}$. The random component is chosen according to a zero mean normal distribution with different standard deviations. In Figure

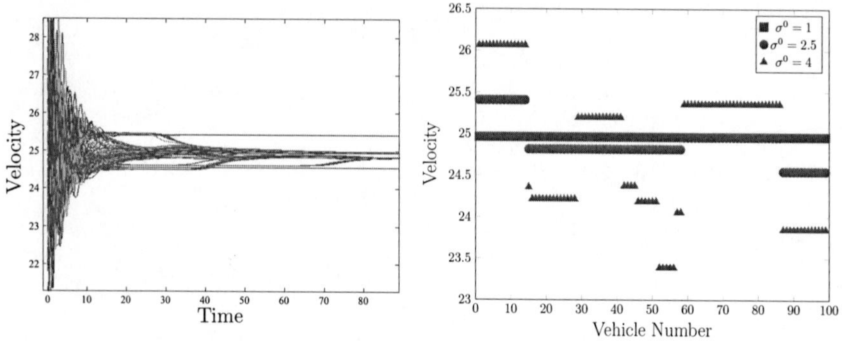

Figure 5.3: Simulation results for a traffic flow model with 100 vehicles placed on a line graph. *Left:* Time trajectories of the velocities for normally distributed coefficients with $\sigma^0 = 2.5$ and $\sigma^1 = 1$. *Right:* Asymptotic velocities predicted by the network optimization problems for $\sigma^0 = 1$ (blue, '□'), $\sigma^0 = 2.5$ (red, 'o'), and $\sigma^0 = 4$ (green, 'Δ').

5.3 (left), the time-trajectories of the velocities v_i are shown with the random coefficients $V_{i,rand}^0, V_{i,rand}^1$ chosen from a distribution with $\sigma^0 = 2.5$ and $\sigma^1 = 1$, respectively. Figure 5.3 (right) shows the asymptotic velocity distribution of the traffic for different choices of the standard deviation σ^0. While for $\sigma^0 = 1$ the traffic agrees on a common velocity, already for $\sigma^0 = 2.5$ a clustering structure of the network can be seen. The clustering structure becomes more refined for $\sigma^0 = 4$. We have chosen for all studies $\sigma^1 = 1$. Please note that the novel network theoretic framework provides us with efficient tools to analyze this non-trivial dynamic behavior.

The Aeyels and De Smet Model: Over the last years, various mathematical models for clustering have been proposed in the literature. We want to discuss here the relation of our analysis to one of the established clustering models. In Aeyels and De Smet (2008), De Smet and Aeyels (2009), a dynamical network model with bounded interaction rules is proposed. The model of Aeyels and De Smet (2008) takes the simple form

$$\dot{\chi}_i = b_i + A_i \sum_{k=(i,j)\in \mathbf{E}} w_k \psi_k(\chi_j - \chi_i), \tag{5.16}$$

with $\psi_k : \mathbb{R} \mapsto [-1, 1]$ being a monotonic bounded function and $A_i > 0$, b_i being constants. It was shown in Aeyels and De Smet (2008), De Smet and Aeyels (2009) that this model exhibits a clustering behavior, and the clustering structure is independent of the initial condition. While the model itself is slightly different from our model (i.e., the model is first order), its clustering behavior can also be characterized using the optimization tools we proposed here.

To understand this connection, we consider the auxiliary second order model

$$\ddot{\chi}_i = -A_i^{-1}(\dot{\chi}_i - b_i) + \sum_{k=(i,j)\in\mathbf{E}} w_k\psi_k(\chi_j - \chi_i). \tag{5.17}$$

It is not difficult to see that the second order model (5.17) falls into the system class we are studying in this chapter. We define the node state as $x_i = \dot{\chi}_i$ to obtain the node dynamics

$$\dot{x}_i = -A_i^{-1}(x_i - b_i) + u_i, \quad y_i = x_i.$$

Clearly, the coupling takes the desired form

$$u_i = \sum_{k=(i,j)\in\mathbf{E}} w_k\psi_k\Big(\int_{t_0}^{t} y_j(\tau) - y_i(\tau)d\tau\Big).$$

It is now easy to see that the clustering structure of the model (5.17) can be analyzed using the optimization framework derived above. The node potential functions for the model are simply the quadratic functions

$$K_i^\star(\mathrm{y}_i) = \frac{1}{2A_i}(\mathrm{y}_i - b_i)^2,$$

and the agreement value is $\beta = \frac{\sum_{i=1}^n A_i^{-1}b_i}{\sum_{i=1}^n A_i^{-1}}$. The corresponding steady state input is $\mathbf{u} = A^{-1}(\beta\mathbb{1} - \mathrm{b})$, where $A = \mathrm{diag}(A_1, \dots, A_n)$ and $\mathrm{b} = [b_1, \dots, b_n]^\top$. The network is clustering if the steady state input $\mathbf{u}$ cannot be represented in the form $\mathbf{u} = E\boldsymbol{\mu}$ with all flows satisfying $|\mu_k| \le w_k$. Then, the exact clustering analysis can be determined by solving the constrained network flow or the saddle-point problem.

As we are interested in the clustering structure of the model of Aeyels and De Smet, we have to show that the clustering structure of (5.17) is identical to the clustering structure of (5.16). In fact, in both models the asymptotic behavior is independent of the initial conditions. From our previous discussion, we know that in the asymptotic behavior of (5.17) the node state will be constant, i.e., $\lim_{t\to\infty} \dot{x}_i(t) = \lim_{t\to\infty} \ddot{\chi}_i \to 0$. Thus, we can restrict the analysis of (5.17) to the asymptotic behavior in the set where $\ddot{x}_i = 0$. Then, the dynamics of the velocity reduces to

$$\begin{aligned} 0 &= -A_i^{-1}(\dot{\chi}_i - b_i) + \sum_{k=(i,j)\in\mathbf{E}} w_k\psi_k(\chi_j - \chi_i) \\ \Leftrightarrow \dot{\chi} &= b_i + A_i \sum_{k=(i,j)\in\mathbf{E}} w_k\psi_k(\chi_j - \chi_i). \end{aligned}$$

Obviously, this is the model (5.16) and both models have the same asymptotic behavior. We can conclude, that the optimization-based clustering analysis method we proposed in this thesis can be used to analyze the clustering structure of the model (5.16).

The Kuramoto Model: The model of Aeyels and De Smet is a variation of a very important dynamic model for synchronization, i.e., the Kuramoto oscillator model. In fact, the Kuramoto oscillator model has a structure that is very similar to the model (5.16), except that the coupling nonlinearities are sinusoidal functions, i.e.,

$$\dot{\theta}_i = b_i + A_i \sum_{k=(i,j)\in \mathbf{E}} w_k \sin(\theta_j - \theta_i). \tag{5.18}$$

Having this observation in mind, it seems natural to ask for the relevance of our analysis method for the classical Kuramoto oscillator model. Research on the synchronization of Kuramoto oscillator models is very actively pursued, see e.g., Strogatz (2000) or the recent survey Dörfler and Bullo (2012). However, it is widely accepted that the exact analysis of synchronization of the Kuramoto model is extremely difficult and up to date no exact analysis method is known. We do not aim to provide here an analysis of the synchronization problem, but we rather want to discuss, how our duality results can be interpreted in the context of oscillator networks.

The first observation we make about the model (5.18) is that the coupling nonlinearities are not monotonic on their complete domain, but only locally. Therefore, Assumption 5.1 is not met, and we cannot expect to obtain an exact clustering result from our analysis. However, we can still hope to obtain some insights. By using the same analysis steps as in the previous example, we can extend the Kuramoto model to a second order model and represent the model in the desired feedback form by introducing the variable $x_i = \dot{\theta}_i$, i.e.,

$$\begin{aligned} \dot{x}_i &= -A_i^{-1}(x_i - b_i) + u_i, \quad y_i = x_i, \\ u_i &= \sum_{k=(i,j)\in \mathbf{E}} w_k \sin\Big(\int_{t_0}^{t} y_j(\tau) - y_i(\tau) d\tau\Big). \end{aligned}$$

Note that the extended model has the same set of equilibrium points as the original model. Thus, for the existence of an equilibrium configuration we can equivalently study the second order model. The agreement value is again $\beta = \frac{\sum_{i=1}^{n} A_i^{-1} b_i}{\sum_{i=1}^{n} A_i^{-1}}$, and the steady state input is $\mathbf{u} = A^{-1}(\beta \mathbf{1} - \mathrm{b})$. We can now use our duality framework to analyze the synchronization problem. Note that the node potential functions are the same as in the previous examples. The difference between the two models is purely in the coupling structure. Therefore, it seems worth to study the potential functions for the edges, that turn out to be

$$P_k(\eta_k) = w_k \left(1 - \cos(\eta_k)\right). \tag{5.19}$$

Note that the functions P_k are *non-convex*. In fact, the functions are only locally convex on the set $\mathcal{C}_k := \{\eta_k \in \mathbb{R} : -\frac{\pi}{2} \le \eta_k \le \frac{\pi}{2}\}$. Anyways, we can formulate the optimal potential problem on the edges, i.e., problem (OPP2), as follows

$$\begin{aligned} \min \quad & \sum_{i=1}^{n} (\beta - b_i) A_i^{-1} \mathrm{v}_i + \sum_{k=1}^{m} w_k \left(1 - \cos(\eta_k)\right) \\ \text{s.t.} \quad & \eta = E^{\top} \mathbf{v}. \end{aligned} \tag{5.20}$$

This optimal potential problem characterizes the existence of a synchronous solution to (5.18).

Proposition 5.13. *The dynamics* (5.18) *has a synchronous solution if and only if the optimization problem* (5.20) *has an optimal solution* $(\boldsymbol{\eta}, \mathbf{v})$ *satisfying* $\boldsymbol{\eta} \in \mathcal{C}_1 \times \cdots \times \mathcal{C}_m$.

Proof. We first proof necessity. If (5.18) has a synchronous solution, then there must exist two static vectors $\mathbf{v}$ and $\boldsymbol{\eta} \in \mathcal{C}_1 \times \cdots \times \mathcal{C}_m$ such that

$$\mathbf{u} = EW \sin(\boldsymbol{\eta}), \; \boldsymbol{\eta} = E^\top \mathbf{v}, \tag{5.21}$$

where $W = \operatorname{diag}(w_1, \ldots, w_m)$, and $\sin(\boldsymbol{\eta}) = [\sin(\eta_1), \ldots, \sin(\eta_m)]^\top$. The vectors $\mathbf{v}$ and $\boldsymbol{\eta}$ satisfy the first order optimality conditions for (5.20) and are therefore optimal solutions. Sufficiency can be proven using exactly the same argument. If there is an optimal solution $\mathbf{v}$ and $\boldsymbol{\eta}$ to (5.20) satisfying $\boldsymbol{\eta} \in \mathcal{C}_1 \times \cdots \times \mathcal{C}_m$, then this pair of vectors satisfies (5.21). Thus, it is an equilibrium point for the dynamics (5.18). □

We can compare now the statements provided by the duality analysis for the two models, the model of Aeyels and De Smet and the Kuramoto model. For both models, the network optimization framework established in this thesis provides a mean to characterize the synchronous steady state behavior of the system. For the model of Aeyels and De Smet the solution of the corresponding network optimization problems characterized the asymptotic behavior of the network exactly. Even if no synchronous solution exists, the network optimization problems still characterize the asymptotic (i.e., clustering) behavior of the model. Additionally, convergence to the predicted asymptotic behavior was independent of the initial conditions of the dynamics. It is remarkable that all four network optimization problems related to the model are *convex*.

Now, for the Kuramoto model only a significantly weaker statement could be made. Still, the synchronous solution, if it exists, could be connected to a locally convex optimization problems. However, a statement concerning the region of attraction of this synchronous solution cannot be easily made. Additionally, the network optimization framework provides no analysis of the behavior of the Kuramoto model, if the synchronous solution does not exist. In this direction, it seems important that the network optimization problems connected to the Kuramoto model are *non-convex*.

Summarizing, we considered the two nonlinear dynamical network models of Aeyels and De Smet and of Kuramoto. It turned out, that the Aeyels and De Smet model can be efficiently analyzed using convex optimization tools, while only fairly weak statements can be made for the Kuramoto model. With this observation in mind, the famous and celebrated statement of R.T. Rockafellars (Rockafellar, 1993) seems now also appropriate for cooperative control problems: *The great watershed in cooperative control is not between linearity and non-linearity, but rather between convexity and non-convexity.*

5.4 Hierarchical Clustering Using a Saddle-Point Analysis

We change in this section our perspective on the clustering problem. Instead of considering the dynamical aspects, we focus now completely on the static clustering problem. Our goal is to understand and to utilize the clustering structures induced by the static optimization problems, in particular by the saddle-point problem (SPP).

In a first step, we aim to derive combinatorial conditions on when clustering will appear. The conceptual idea we will exploit is to consider the network as a weighted graph. The node weights will be the node divergences resulting from the optimal flow problem without capacity constraints. This corresponds directly to the gradient of the node potential functions, i.e., the equilibrium input-output map of the underlying dynamical system, evaluated at the "agreement" solution. The weights on the graph edges are simply the corresponding capacities. Based on this interpretation, we will provide a fully combinatorial condition for clustering of the network. We show that clustering appears, if there are partitions of the network in which the sum of node weights in one partition exceeds the summed capacities of the corresponding cut set. This result can be easily understood, when the flow network interpretation is considered. To support this intuition, we show that the clustering condition results from the max-flow/min-cut duality relation.

The presented discussions give an in depth understanding of the clustering problem. Most important, they lead to an algorithmic procedure for the identification of hierarchical clustering structures in weighted networks. The proposed method identifies clustering structures by solving repeatedly convex optimization problems. We illustrate the usefulness of the method on the structural analysis of large scale power networks and show that our method generalizes a know computational procedure and efficiently computes hierarchical clustering structures.

5.4.1 Combinatorial Conditions for Clustering

We investigate in this section the combinatorial aspects of the clustering problem. We will keep the dynamical system's interpretation in the back of our minds, but we will work here only with the static optimization problems. A first contribution we make is the proposal of a measure, which indicates whether a network is synchronizing or clustering. The idea behind the coefficient is simply to test, based on the solution of the network optimization problems without flow constraints (OFP1), (OPP1), whether the given capacities are sufficiently large or cause a clustering of the network.

Let $\bar{\mathbf{y}}$ and $\bar{\mathbf{u}}$ be solutions to (OPP1) and (OFP1), respectively, and define $W = \operatorname{diag}\{w_1, \dots, w_m\}$. In the flow network interpretation, $\bar{\mathbf{u}}$ will be the optimal divergence (i.e., in-flow) of the network, and W is the diagonal matrix of the transportation capacities. In the dynamical systems interpretation, $\bar{\mathbf{y}} \in \operatorname{span}\{\mathbf{1}\}$ is the agreement steady state. Thus, $(\bar{\mathbf{u}}, \bar{\mathbf{y}})$ is the steady state input-output pair corresponding to output agreement. The matrix W represents the "coupling strength" between the dynamical systems.

Definition 5.14. *The synchronization coefficient γ^* is*

$$\begin{aligned} \gamma^* := \min_{\boldsymbol{\mu}} \ & \|W^{-1}\boldsymbol{\mu}\|_\infty \\ \text{s.t. } & \bar{\mathbf{u}} + E\boldsymbol{\mu} = 0. \end{aligned} \tag{5.22}$$

The value γ^* contains important information on the synchronization or clustering structure of the network. Keeping the flow network interpretation in mind, one can directly see that the network is synchronizing if $\gamma^* \leq 1$ and is clustering if $\gamma^* > 1$. An important observation is that the value γ^* is always determined by a cut-set of $\mathcal{G}$.

Proposition 5.15. *Let $\bar{\boldsymbol{\mu}}$ be a solution to (5.22) and let $\mathbf{Q} \subset \mathbf{E}$ be the set of edges for which $|\mu_k^*| = \gamma^* w_k$. Then $\mathbf{Q}$ is a cut-set.*

Proof. Given any solution $\bar{\boldsymbol{\mu}}$ satisfying the equality constraint of (5.22). One can always find a $\tilde{\boldsymbol{\mu}}$ as a variation of $\bar{\boldsymbol{\mu}}$ in the flow-space of $\mathcal{G}$ such that for a particular edge $k \in \mathbf{E}$, contained in a cycle $\mathcal{C}$, $|\tilde{\mu}_k| < |\bar{\mu}_k|$. However, there must then be at least one other edge $l \in \mathbf{E}$ in the cycle $\mathcal{C}$ for which $|\tilde{\mu}_l| > |\bar{\mu}_l|$. The result is now demonstrated via contradiction. Assume that the set of edges $\mathbf{Q}$, with $|\bar{\mu}_k| = \gamma^* w_k$ for $k \in \mathbf{Q}$, does not form a cut-set. Then every edge $k \in \mathbf{Q}$ must be contained in at least one cycle of $\mathcal{G}$, say $\mathcal{C}$, such that $(\mathcal{C} \setminus \{k\}) \cap \mathcal{Q} = \emptyset$. This, however, implies that one can find a δ, with $|\delta|$ sufficiently small, and define $\tilde{\boldsymbol{\mu}} = \bar{\boldsymbol{\mu}} + \delta s$, where s is the signed path vector corresponding to the cycle $\mathcal{C}$, such that $|\tilde{\mu}_k| < |\bar{\mu}_k| = \gamma^* w_k$ for $k \in \mathbf{Q}$ and $|\tilde{\mu}_l| < \gamma^* w_l$ for all other edges l in $\mathcal{C}$. This contradicts the original assumption that $\gamma^* = \min \|W^{-1}\boldsymbol{\mu}\|$. □

We will refer to $\mathbf{Q}$ in the following as the γ^*-*cut set*. These quantities also have a combinatorial interpretation that we explore next. We consider in the following only bi-partitions of the (weighted) graph $\mathcal{G}$. The set of all possible bi-partitions of $\mathcal{G}$ is denoted by $\mathbb{P}_2(\mathcal{G})$. A particular bi-partition $\mathbb{P} \in \mathbb{P}_2(\mathcal{G})$ is then characterized by the triplet $\mathbb{P} = (\mathbf{P}_1, \mathbf{P}_2, \mathbf{Q})$, where $\mathbf{P}_1$ and $\mathbf{P}_2$ are disjoint node sets, such that $\mathbf{P}_1 \cup \mathbf{P}_2 = \mathbf{V}$ and $\mathbf{Q} \subset \mathbf{E}$ is the set of edges connecting the two node sets. We define the *quality* of a bi-partition in the following.

Definition 5.16. *The* quality *of a bi-partition $\mathbb{P} = (\mathbf{P}_1, \mathbf{P}_2, \mathbf{Q}) \in \mathbb{P}_2(\mathcal{G})$ is*

$$\Psi(\mathbb{P}) = \frac{|\sum_{i \in \mathbf{P}_1} \bar{\mathrm{u}}_i|}{\sum_{k \in \mathbf{Q}} w_k} = \frac{|\sum_{i \in \mathbf{P}_1} \nabla K_i^*(\bar{\mathrm{y}}_i)|}{\sum_{k \in \mathbf{Q}} w_k}. \tag{5.23}$$

The concept of the cut quality becomes meaningful, when we consider the network as a static weighted graph. Each node in the graph is weighted with the optimal divergence $\bar{\mathrm{u}}_i$, and each edge with w_k. Now, the quality of a graph partition is a purely combinatorial property. Recall that $\sum_{i=1}^n \bar{\mathrm{u}}_i = 0$ and therefore the quantity $|\sum_{i \in \mathbf{P}_1} \bar{\mathrm{u}}_i|$ can be interpreted as the weighted *imbalance* of the two clusters. Thus, the quality $\Psi(\mathbb{P})$ measures the ratio of the weight imbalance and the capacity of the corresponding cut-set $\sum_{k \in \mathbf{Q}} w_k$. The next result, which has the flavor of a duality relation, explains the connection between the synchronization coefficient and the quality of the cut-sets.

Theorem 5.17. *Let $\mathbb{P}_2(\mathcal{G})$ be the set of all possible bi-partitions of $\mathcal{G}$, then*

$$\max_{\mathbb{P} \in \mathbb{P}_2(\mathcal{G})} \Psi(\mathbb{P}) = \min_{\substack{\boldsymbol{\mu} \\ \bar{\mathrm{u}} + E\boldsymbol{\mu}}} \|W^{-1}\boldsymbol{\mu}\|_\infty \tag{5.24}$$

Proof. The statement is equivalent to $\max_{\mathbb{P} \in \mathbb{P}_2(\mathcal{G})} \Psi(\mathbb{P}) = \gamma^*$. First, we show that $\Psi(\mathbb{P}) \leq \gamma^*$ for all $\mathbb{P} \in \mathbb{P}_2(\mathcal{G})$. Given γ^*, there exists a solution to (5.22) $\bar{\boldsymbol{\mu}}$ satisfying $|\bar{\mu}_k| \leq w_k \gamma^*$ for all k. One can now choose any bi-partition $\mathbb{P} = (\mathbf{P}_1, \mathbf{P}_2, \mathbf{Q}) \in \mathbb{P}_2(\mathcal{G})$, and define, without loss of generality, $\mathbf{P}_1$ to be such that $\sum_{i\in\mathbf{P}_1} \bar{\mathrm{u}}_i > 0$. For a bi-partition, we define an *indicator vector* $\zeta \in \{-1,1\}^n$ such that $\zeta_i = +1$ if node $i \in \mathbf{P}_1$ and $\zeta_i = -1$ if node $i \in \mathbf{P}_2$. Note that $\zeta^\top \bar{\mathbf{u}} = 2\sum_{i\in\mathbf{P}_1} \bar{\mathrm{u}}_i$. Multiplying the feasibility condition $\bar{\mathbf{u}} + E\bar{\boldsymbol{\mu}} = 0$ from the left with the indicator vector leads to

$$\zeta^\top \mathbf{u} + \zeta^\top E\bar{\boldsymbol{\mu}} = 2\sum_{i\in\mathbf{P}_1} \mathrm{u}_i + \zeta^\top E\bar{\boldsymbol{\mu}} = 0. \tag{5.25}$$

Given the indicator vector of a partition, define a new vector $c = \frac{1}{2}E^\top\zeta$, which has a very characteristic structure

$$c_k = \begin{cases} +1 & \text{if edge } k \text{ originates in } \mathbf{P}_1 \\ -1 & \text{if edge } k \text{ terminates in } \mathbf{P}_1 \\ 0 & \text{if edge } k \notin \mathbf{Q}. \end{cases}$$

Now, the condition (5.25) can be written as $\sum_{i\in\mathbf{P}_1} \mathrm{u}_i = -c^\top\bar{\boldsymbol{\mu}}$. Since $|\bar{\mu}_k| \le w_k\gamma^*$ for all k, we obtain the upper bound $-c^\top\bar{\boldsymbol{\mu}} \le \sum_{k\in Q} w_k\gamma^*$. This bound leads to $\sum_{i\in\mathbb{P}_1} \bar{\mathrm{u}}_i \le \sum_{k\in Q} w_k\gamma^*$ and to the conclusion $\Psi(\mathbb{P}) \le \gamma^*$. This last inequality has to hold for any possible bi-partition, proving the first direction.

In the second step, we show that $\max_{\mathbb{P}\in\mathbb{P}_2(\mathcal{G})} \Psi(\mathbb{P}) \ge \gamma^*$. With the goal to arrive at a contradiction, we assume that $\max_{\mathbb{P}\in\mathbb{P}_2(\mathcal{G})} \Psi(\mathbb{P}) < \gamma^*$. It follows from the assumption that

$$\frac{\sum_{i\in\mathbf{P}_1} \mathrm{u}_i}{\sum_{k\in\mathbf{Q}} w_k} < \gamma^* \Leftrightarrow \sum_{i\in\mathbf{P}_1} \mathrm{u}_i < \gamma^* \sum_{k\in\mathbf{Q}} w_k \tag{5.26}$$

for any possible bi-partition. We define, as in the first part of the proof, an indicator vector ζ for any bi-partition. Using the same argumentation as before, we conclude

$$\sum_{i\in\mathbf{P}_1} \mathrm{u}_i = -c^\top\bar{\boldsymbol{\mu}}. \tag{5.27}$$

Combining the two conditions (5.26) and (5.27) leads to the new condition

$$-c^\top\mu^* < \gamma^* \sum_{k\in\mathbf{Q}} w_k; \tag{5.28}$$

this must hold for any possible bi-partition. As a consequence of (5.28), there cannot be a bi-partition $\mathbb{P}$ with a cut-set $\mathbf{Q}$ such that $\bar{\mu}_k = -c_k w_k\gamma^*$ for all $k \in \mathbf{Q}$. However, if there is no such cut-set $\mathbf{Q}$, then every edge $l \in \mathbf{E}$ for which $|\bar{\mu}_l| = w_l\gamma^*$ must be contained in at least one cycle $\mathbf{C}$ for which every other edge $s \in \mathbf{C}\setminus\{l\}$ has $|\bar{\mu}_s| < w_s\gamma^*$. But this implies now that one can find another $\tilde{\boldsymbol{\mu}}$, satisfying all constraints such that $|\tilde{\mu}_l| < w_l\gamma^*$ for every edge $l \in \mathbf{C}$. This contradicts the definition of γ^*, as being the minimal value satisfying $\gamma^* \le \frac{|\bar{\mu}_k|}{w_k}$ for all edges. This is the contradiction we were looking for, and we can conclude that $\max_{\mathbb{P}\in\mathbb{P}_2(\mathcal{G})} \Psi(\mathbb{P}) \ge \gamma^*$. We conclude that $\max_{\mathbb{P}\in\mathbb{P}_2(\mathcal{G})} \Psi(\mathbb{P}) = \gamma^*$. □

We can directly give the following combinatorial condition for clustering.

Corollary 5.18. *The network is clustering if and only if there exists a bi-partition $\mathbb{P} \in \mathbb{P}_2(\mathcal{G})$ such that*

$$\Psi(\mathbb{P}) > 1.$$

The previous discussion allows us to establish a direct connection to a very classical network optimization problem. In fact, for certain choices of the node objective functions, the γ^* cut-set solves the min s-t-cut problem of Ford and Fulkerson (1956). This result will directly match to intuition one might have by considering the flow network interpretation. However, it seems worthwhile to present this fundamental duality result here. In the following, we say that the capacity of a cut $\mathbf{Q}$ is the sum of the edge capacities $\sum_{k\in\mathbf{Q}} w_k$.

Definition 5.19 (Min-s-t cut). *Given a sink node $s \in \mathbf{V}$ and a source node $t \in \mathbf{V}$, the min s-t-cut is the cut $\mathbf{Q}$ that separates s and t and has minimal capacity.*

Choose now the node objective functions in the following way. Pick a constant $\xi \in \mathbb{R}_{>0}$ and set

$$K_i^\star(\mathrm{y}_i) = \begin{cases} \frac{1}{2}(\mathrm{y}_i - \xi)^2 & \text{if } i = s \\ \frac{1}{2}(\mathrm{y}_i + \xi)^2 & \text{if } i = t \\ \frac{1}{2}\mathrm{y}_i^2 & \text{if } i \neq s, t. \end{cases} \tag{5.29}$$

Lemma 5.20. *Consider a network $\mathcal{G}$ with node potential functions chosen as* (5.29)*. Let $\mathbf{Q}_{st,1}, \ldots, \mathbf{Q}_{st,q}$ be all minimal s-t-cuts. Then the γ^* cut-set is the union of* all *min s-t-cuts.*

Proof. It can be easily seen that $\bar{\mathbf{y}} = 0$ and consequently $\bar{\mathrm{u}}_s = \nabla K_s^\star(0) = -b$, $\bar{\mathrm{u}}_t = \nabla K_t^\star(0) = b$ and $\bar{\mathrm{u}}_i = \nabla K_i^\star(0) = 0$. Thus, for any bi-partition $\mathbb{P} = (\mathbf{P}_1, \mathbf{P}_2)$ such that $s \in \mathbf{P}_1$ and $t \in \mathbf{P}_2$, it holds that $|\sum_{i\in\mathcal{P}_1} \mathrm{u}_i| = b$. For any partition not separating s and t holds $|\sum_{i\in\mathcal{P}_1} \mathrm{u}_i| = 0$. The γ^* cut-set corresponds to the partition with maximal quality, which takes here the value $\Psi(\mathbb{P}) = \frac{b}{\sum_{k\in\mathbf{Q}} w_k}$. Now, the statement follows immediately. □

This is fully in accordance with the network flow interpretation. However, we want to remark that the connection is not that obvious for the passivity-based cooperative control framework. As we have now a first combinatorial condition for clustering of a weighted graph, we might ask how this idea can be used for a further network analysis. In fact, considering all previous discussions and the examples of the previous section, we can observe that the clustering behavior can be used as a computational tool for finding clustering structures in weighted networks. We formalize the idea in the next section.

5.4.2 A Hierarchical Clustering Algorithm

Up to now, we have taken only an analytic perspective. That is, we have assumed that a certain dynamical network configuration is given and we have developed a machinery to predict the corresponding clustering structure. The proposed analysis relies heavily on convex optimization, while still explicit combinatorial interpretations exist. An immediate question following the previous analysis is, whether the developed machinery can be used to obtain more structural information about a network. The question is: *can we use the concept of cluster synchronization to define, whether groups of nodes are strongly or weakly connected?* Such problems are known as (static) *community detection* problems in graphs (Fortunato, 2010). Roughly speaking, we might say that two nodes are strongly connected, if they synchronize "easily" while they are weakly connected if they partition "easily". Of course, we need to define more precisely, what it means that nodes are synchronizing or partitioning "easily".

In the following, we assume that a network $\mathcal{G}$ is given and each node $i \in \mathbf{V}$ is assigned a strictly convex cost function $K_i^\star(\mathrm{y}_i)$. We want to point out that for the discussion presented here the functions $K_i^\star(\mathrm{y}_i)$ need not necessarily to result from the input-output maps of dynamical systems, but can also be chosen as general strongly convex loss or penalty functions. Assume, for example, that each node is assigned a parameter or weight $\xi_i \in \mathbb{R}$, then the node cost function might be simply chosen as a quadratic loss function, i.e.,

$$K_i^\star(\mathrm{y}_i) = \frac{1}{2}(\mathrm{y}_i - \xi_i)^2.$$

We assume additionally that all edges in the network have the same capacity, i.e., $w_k = w$, for $k \in \{1, \dots, m\}$. The advantage of considering identical edges is that the capacity becomes a single parameter for the network. We will use a variation of this parameter to uncover a hierarchical clustering structure in the weighted networks.

The first observation we make is that there exists $w_{\min}$, called the *minimal agreement capacity*, such that the network reaches full agreement for all $w \geq w_{\min}$. It can be easily seen that $w_{\min}$ can be computed with the linear program (5.22) by simply choosing $W = I$. Now, analogous to the γ^* cut-set, we introduce the idea of the first-cut.

Definition 5.21. *The set of edges $\mathbf{Q}_{fc} \subset \mathbf{E}$ that are saturated for $w = w_{\min}$ are called the* first-cut.

The name "first-cut" emphasizes that this is the first set of edges to become saturated as w is decreased from a large value. The first-cut is just the γ^* cut-set for a network with identical capacities. Consequently, the combinatorial interpretation of the first cut follows as a direct consequence of the proof for Theorem 5.17. Assume the first-cut $\mathbf{Q}_{fc}$ induces a bi-partition $\mathbb{P}_{fc}$, then $\mathbb{P}_{fc} = \arg\max_{\mathbb{P} \in \mathbb{P}^2(\mathcal{G})} \Psi(\mathbb{P})$. Additionally, we can characterize the first-cut $\mathbf{Q}_{fc}$ precisely in terms of the quality of all partitions in a network.

Corollary 5.22. *Assume $\mathbb{P}_1, \dots, \mathbb{P}_q$ are all bi-partitions maximizing $\Psi(\mathbb{P})$ and let $\mathbf{Q}_1, \dots, \mathbf{Q}_q$ be the corresponding cut-sets. Then the first-cut is given by $\mathbf{Q}_{fc} := \mathbf{Q}_1 \cup \dots \cup \mathbf{Q}_q$.*

The first-cut gives us a straight forward way to partition a network, according to a criterion that can be easily understood in the context of flow networks. Given a certain and fixed in-flow to a network, $w_{\min}$ is the smallest capacity required to transport the flow through the network. The first-cut is accordingly defined as the set of edges carrying the most flow. While this interpretation is intuitive for flow networks, it is less obvious in the context of dynamical systems, where $w_{\min}$ represents the minimal coupling "strength" required for synchronization.

The first-cut is a meaningful way to find a first partitioning of the network. However, it does not give us the complete structural information about the network. In fact, this analysis does not tell us, whether the partitions we find are in itself strongly connected or not, and provides therefore no information about how meaningful the partitioning is. In fact, it might be preferable to consider a further partitioning. We can obtain this information from our cluster synchronization analysis. The question we ask is simply, whether the nodes will still reach an agreement if we further decrease the edge capacity. This question seems obvious if we think about cooperative dynamical systems. However, there is no equivalent interpretation when thinking about flow networks. Therefore, only considering clustering in dynamical networks opens the door for a hierarchical clustering analysis.

We denote again the *set of all possible p-partitions* that can be formed in the graph $\mathcal{G}$ by $\mathbb{P}^p(\mathcal{G})$. Partitions of $\mathcal{G}$ are sometimes ordered in a *hierarchical* manner, with smaller partitions being contained in the larger partitions.

Definition 5.23. *A partition $\mathbb{P}_i \in \mathbb{P}^k(\mathcal{G})$ is a* successor *of $\mathbb{P}_j \in \mathbb{P}^l(\mathcal{G})$ with $k > l$, i.e., $\mathbb{P}_i \succ \mathbb{P}_j$, if $\mathbb{P}_j$ can be formed by merging components from $\mathbb{P}_i$. We write $\mathbb{P}_i \succeq \mathbb{P}_j$ if $\mathbb{P}_i$ is either a successor or exactly identical to $\mathbb{P}_j$.*

We first observe the following result, that follows directly from the convexity of the underlying optimization problems.

Lemma 5.24. *For any two scalars $w_1 > w_2$, let $\mathbb{P}_{w_1}$ and $\mathbb{P}_{w_2}$ be the partitions induced by the solution of the saddle-point problem (SPP). Then it holds that $\mathbb{P}_{w_2} \succeq \mathbb{P}_{w_1}$.*

Based on this observation, we derive a simple method to detect the hierarchical clustering structure. The algorithmic idea is described in Algorithm 4.

Algorithm 4 Hierarchical Clustering

Require: $K_i^\star(\mathrm{y}_i)$, E, $T \in \mathbb{Z}_{>0}$;
  Compute $w_{\min}$;
  for $\tau = 0, \ldots, T-1$ **do**
    Set $w(\tau) = w_{\min}(1 - \frac{\tau}{T})$;
    Solve

$$\min_{\mathbf{y}} \sum_{i=1}^{n} K_i^\star(\mathrm{y}_i) + w(\tau) \cdot \|E^\top \mathbf{y}\|_1; \tag{5.30}$$

    Set $\mathbb{P}(\tau)$ as partition induced by $\mathbf{y}$;
  end for

This simple algorithm requires the knowledge of node objective functions $K_i^\star(\mathrm{y}_i)$, the graph incidence matrix E and the number of discretization steps T. Then it computes the minimal agreement capacity $w_{\min}$ by solving, e.g., a linear program of the form (5.22). As we know that the systems will reach an agreement for all $w \geq w_{\min}$, we do not have to consider these capacities. Starting from $w = w_{\min}$ we repeatedly reduce the edge capacity and solve the convex optimization problem (5.30). This problem is exactly identical to (OPP3), but takes a very characteristic structure here. Since all edge capacities are identical, the objective functions associated to the tensions, i.e., $\boldsymbol{\zeta} = E^\top \mathbf{y}$, become simply the ℓ_1 norm. If we interpret $K_i^\star$ as a loss function, the clustering problem is the loss minimization plus an ℓ_1-norm regularization term. It seems worth to point out that the optimization problem is structurally very close to the well-known Lasso problem (Tibshirani, 1996), if the node cost functions are quadratic. However, in difference to the classical Lasso problem, the clustering problem penalizes not the variables $\mathbf{y}$ directly, but rather the differences of the variables, i.e., the tensions $\boldsymbol{\zeta}$.

5.4.3 Application Example: Structural Analysis of Power Networks

Dynamic clustering is a well-known instability phenomenon in power networks, where in the case of failures the network often partitions into groups. Generators within one group synchronize their frequencies but they are out of synchrony to generators in other groups. Such a behavior was, e.g., observed in the 2006 power blackout, where the European power grid partitioned into three areas, see for example Union for the Coordination of transmission of Electricity (UCTE) (2007). It remains a difficult problem to relate the graph topology and the expected clustering structure of power networks. In particular in large scale power networks with several thousands of nodes and connections, very efficient computational methods are required to identify structures in the network.

From a computational perspective, the optimal partitioning analysis of power systems has attained significant attention. Most existing work focuses on the detection of critical cut-sets (or optimal partitions) within power networks, see e.g., Lesieutre et al. (2006), Bompard et al. (2010). These methods intend to compute one partition of the network, but not an hierarchical structuring. One of these approaches, particularly interesting in the context of this work, is the *inhibiting bisection problem* studied in Pinar et al. (2006) and Lesieutre et al. (2006). We will briefly review the problem and the according solution approach here.

Each network node is assigned a weight p_i representing either power generation ($p_i > 0$) or power consumption ($p_i < 0$). It is assumed that the power is balanced over the complete network, that is $\sum_{i=1}^n p_i = 0$. The inhibiting bisection problem is now to find a bipartition of the network that (i) has the minimal cut size and (ii) leads to the maximal power imbalance between the two clusters. We review here the mathematical problem representation studied in Lesieutre et al. (2006). Therefore, we need the notion of *indicator vectors.* A vector $\xi \in \{-1, 1\}^n$ is said to be an indicator vector for a bi-partition $\mathbb{P} = (\mathbf{P}_1, \mathbf{P}_2)$ if $\xi_i = 1$ if $i \in \mathbf{P}_1$ and $\xi_i = -1$ if $i \in \mathbf{P}_2$. The set of indicator vectors for all possible bi-partitions is denoted by $\mathbb{I}_2$. Note that for a given bi-partition $\mathbb{P}$ with the corresponding indicator vector $\xi \in \mathbb{I}_2$, the size of the cut-set is $|\mathbf{Q}| = \frac{1}{4}\xi^\top L\xi$, where $L = EE^\top$ is the Laplacian of $\mathcal{G}$. Additionally, given the vector of power imbalances $\mathbf{p} = [p_1, \ldots, p_n]^\top$, the power imbalance between the partitions computes as $|\mathbf{p}^\top\xi| = 2|\sum_{i\in\mathbf{P}_1} p_i|$. Now, in Lesieutre et al. (2006) the inhibiting bisection problem is formalized as follows

$$\min_{\xi\in\mathbb{I}_2} \quad \xi^\top L\xi + \delta|\mathbf{p}^\top\xi|, \tag{5.31}$$

for some weighting factor $\delta > 0$. This is a non-convex combinatorial problem. However, we show that the problem is closely related to our approach.

Lemma 5.25. *Let the node objective function be* $K_i^\star(y_i) = \frac{1}{2}(y_i - p_i)^2$ *with* $\sum_{i=1}^n p_i = 0$. *Assume that the first-cut induces a bi-partition and let* ξ^* *be the indicator vector of this bi-partition. Then* ξ^* *is a minimizer of*

$$\min_{\xi\in\mathbb{I}_2} \quad \xi^\top L\xi - \left(\frac{2}{w_{\min}}\right)|\mathbf{p}^\top\xi|, \tag{5.32}$$

where $w_{\min}$ *is the minimal agreement capacity and* $\mathbf{p} = [p_1, \ldots, p_n]^\top$.

Proof. With the given node objective functions, it follows directly $\mathbf{u} = -\mathbf{p}$. Now, the bi-partition with highest quality corresponds to the optimal solution to

$$\max_{\xi\in\mathbb{I}_2} \ 2(|\mathbf{p}^\top\xi|)(\xi^\top L\xi)^{-1}.$$

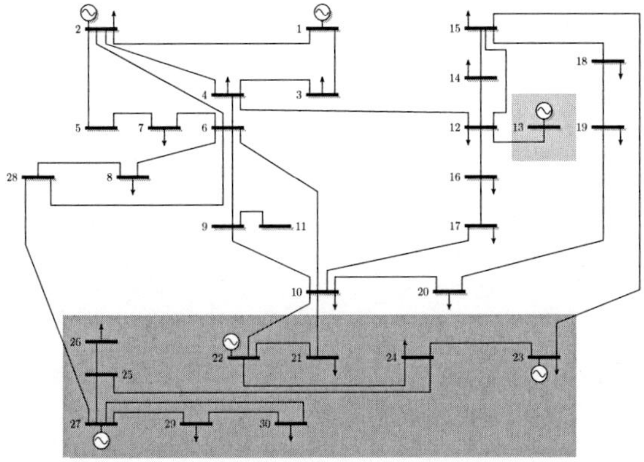

Figure 5.4: Network topology of the IEEE 30-bus network.

The value of the optimal bi-partition (which is induced by the first-cut) is $w_{\min}$. Thus, the optimal indicator vector ξ^* must be such that $w_{\min} = 2(|\mathbf{p}^\top \xi|)(\xi^\top L \xi)^{-1}$. The problem to solve is therefore the feasibility problem of finding $\xi \in \mathbb{I}_2$ such that $2|\mathbf{p}^\top \xi| - w_{\min} \xi^\top L \xi = 0$. By assumption, the first cut is unique. It follows now from the definition of the first cut that equality holds only for the indicator vector corresponding to the first-cut, and is strictly negative for any other vector. Therefore, finding the indicator corresponding to the first cut is equivalent to solving (5.32). □

This discussion shows that the first partition computed by our hierarchical clustering method is an instant of the inhibiting bisection problem. Please note that while problem (5.32) has integer constraints, the optimization problems we are solving are convex. We present now a computational study to illustrate the applicability of our results. In the first step, we use the same problem setup as considered in Lesieutre et al. (2006). In particular, we consider a modified version of the IEEE 30-bus power network. The network topology of the IEEE 30 Bus network is reported in (Power Systems Test Case Archive,

Bus i	p_i [MW]	Bus i	p_i [MW]	Bus i	p_i [MW]
1	145.3	11	0	21	-175
2	-7.3	12	-112	22	315.9
3	-24	13	410	23	310
4	-76	14	-62	24	-87
5	0	15	-82	25	0
6	0	16	-35	26	-35
7	-228	17	-90	27	469.1
8	-300	18	-32	28	0
9	0	19	-95	29	-24
10	-58	20	-22	30	-106

Table 5.1: Network data of the IEEE 30-bus system.

2013). Each node represents a bus in the system. Each bus is connected to a generator, a load or simply to another bus. The power "injection" p_i at each node is the difference of generated and consumed power. The network topology and the considered power injections are summarized in Figure 5.4 and Table 5.1, respectively. In Lesieutre et al. (2006) the inhibiting bisection problem showed that nodes $\{22, 23, 24, 25, 26, 27, 29, 30\}$ as well as the single node $\{13\}$ are loosely connected to the rest of the network. We use now our hierarchical clustering algorithm to detect the clustering structure. Figure 5.5 visualizes how the optimal solutions to the problem (5.30) vary as a function of w. Observe the characteristic *dendrogram* structure of the function illustrated in Figure 5.5. For very large values of w the entire network is in agreement, while as w decreases a partitioning of the network appears (represented by multiple values for y_i). This provides clearly a *hierarchical clustering structure* of the network. A deeper analysis of the clustering structure reveals that the first-cut partitions only a single node, node 13 (marked by the yellow box in Figure 5.4). However, the second partition that appears separates the network nodes $\{22, 23, 24, 25, 26, 27, 29, 30\}$ (illustrated by the green box in Figure 5.4). Our analysis method detects the same partition as the inhibiting bisection problem, but also provides a *complete hierarchical clustering structure.*

5.5 Conclusions

We studied in this section the phenomenon of clustering in dynamical networks. We started the discussion by considering a certain class of static optimal flow problems with capacity constraints on the edges. Duality theory led us then the way to see that the variables of the dual problem, i.e., the optimal potential problem, have a clustered structure, in the sense that there are groups of nodes in the network with the same potential. To see this, we considered a certain instance of the duality relation. Instead of considering purely the

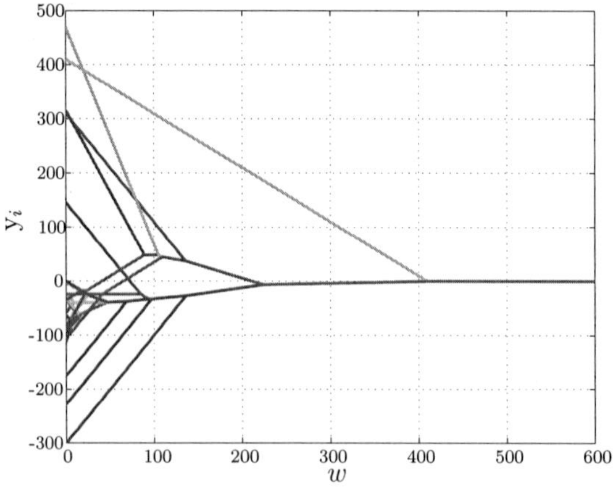

Figure 5.5: Optimal potential variables y_i as function of the edge capacities w.

optimal flow problem or purely the optimal potential problem, we studied the saddle-point problem that is "between" these two dual problems. The saddle-point problem involves directly the potential and flow variables and leads the way to a precise clustering analysis.

Following this, we turned our attention to a class of dynamical networks that exhibit clustering in their asymptotic behavior. Therefore, we established a connection between the constrained optimal flow problems and networks of equilibrium independent dynamical systems with saturated nonlinear couplings. We showed that the asymptotic behavior of the dynamic network can be connected to the solution of the static optimization problem. This observation directly provided a proof of the clustering behavior, along with a computationally attractive analysis tool. We used a Lyapunov analysis to prove the convergence of the dynamical network to the clustered solution structure.

Finally, we used the idea of cluster synchronization to identify hierarchical clustering structures in static and dynamic networks. We used the capacity bounds on the flow variables as parameters and showed that a variation of this parameter reveals a hierarchical clustering structure. Using this idea, we were able to present a combinatorial condition for the clustering behavior. Additionally, we could establish an intimate connection to the classical min s-t-cut problem. For identifying the hierarchical clustering structures, we proposed then a simple algorithm, relying on the sequential solution of a convex optimization problem. Furthermore, we illustrated the applicability of the method on a benchmark network partitioning problem.

We can summarize now the broad framework established in last two chapters. We have developed a duality framework for passivity-based cooperative control problems. The identified duality relations have several practical implications. In particular, the duality framework and the corresponding optimization problems turned out to be efficient tools for analyzing the emergent behavior of clustering. Our analysis presented in the last two chapters suggests that a distinction between "easy" and "difficult" cooperative control problems is not related to linearity or non-linearity, but rather to convexity or non-convexity.

Chapter 6

Conclusions and Outlook

6.1 Conclusions

Cooperative optimization and control are enabling technologies for modern engineering systems. In fact, we have presented throughout the thesis various engineering problems that justify this broad claim. We have discussed, for example, a number of relevant decision problems, ranging from estimation in sensor networks to the control of microgrids, that inherently require to solve optimization problems in a distributed way. Additionally, different dynamical networks, such as inventory or traffic networks, were shown to require a design that directly ensures an optimal operation. This thesis discussed on the one hand the relevance of approximation methods in cooperative optimization and on the other hand the relevance of duality and convexity in cooperative control.

A contribution is made in this thesis towards answering the question, how can decisions be taken efficiently in a distributed environment. We considered a general formulation of distributed convex optimization problems, where processors are assigned convex constraint sets and have to reach an agreement on the optimizer of a linear cost function over the intersection of these sets. We relied in this thesis a powerful conceptual idea for distributed optimization: the use of *approximation methods.* Instead of approaching the distributed optimization problem directly, we proposed here to solve repeatedly approximations of the original problem. Utilizing this conceptual idea, only a simple and well-structured approximated problem had to be solved in a distributed way. Of course, this is significantly easier than solving the original problem in a distributed way. This thesis focused on polyhedral approximation methods that use linear programs as approximations. From the polyhedral approximation idea it follows directly that it is sufficient to solve linear programs in a distributed way in order to solve general convex optimization problems. We proposed a simple but efficient algorithmic idea for distributed linear programming. Processors store and exchange a small number of (linear) constraints, namely the basis constraints. By repeatedly exchanging their basis constraints and by solving repeatedly small linear programs, the processor network computes cooperatively the optimal solution to the linear program. We have shown that this fundamental algorithmic idea translates directly into a general framework for distributed convex optimization, where processors have to repeatedly solve and refine the approximation of the original program. In view of this, the proposed approximation method provides a broad framework for distributed optimization.

We have presented in this thesis several variants of the approximation idea for solving different classes of optimization problems. For example, nominal convex optimization problems defined by inequality or semidefinite constraints can be solved by using subgradients to define the linear approximations. Additionally, we have shown that robust optimization problems can be solved by using a certain pessimizing oracle. We have then presented the somehow surprising result that trajectory exchange methods, as they are widely studied in the context of distributed model predictive control, can be understood as polyhedral approximation methods, when their dual formulation is considered. Based on the general algorithmic idea proposed in this thesis, we then presented a novel version of trajectory exchange method that can solve the finite horizon optimal control problems underlying cooperative predictive control to optimality. To derive these conclusions, we exploited the duality between the original predictive control problem and a distributed convex optimization problem with semi-infinite constraints.

Following the discussion of approximation methods in cooperative optimization, where duality played already an important role, we showed that duality is also a concept of major importance in cooperative control. A main contribution of this thesis was the identification of several duality relations between the dynamic variables in output agreement problems. In particular, we showed that the output variables in cooperative control have an interpretation as potential variables in the network theoretic sense defined by Rockafellar (1998). In this network theoretic sense, the duals of potentials are divergences. We have shown that the input variables of the cooperative control problem have an interpretation as divergences and are thus dual variables to the output variables. This duality interpretation followed directly, after we associated the asymptotic behavior of the dynamical network to several network optimization problems. Beyond the theoretical elegance, the duality relations turned out to be of significant practical relevance. The duality relations and the associated network optimization problems answer the fundamental question, how cooperative dynamical systems have to be designed to work optimally together. We illustrated this finding on an inventory control problem.

Beyond these direct implications, the duality relations also enabled us to study more complex phenomena in dynamic networks. As a main contribution of this thesis, we have connected the clustering phenomenon in dynamical networks to the duality framework. The duality framework led directly to computational methods for analyzing the clustering behavior. While studying the clustering phenomenon, we compared a dynamical model for clustering, as studied in this thesis, to the well-known Kuramoto model and encountered thereby a fundamental property of cooperative control. We observed that the boundary between problems that can be exactly analyzed and those that are hard to analyze is not the linearity or non-linearity of the models, but rather the convexity or non-convexity of the associated problems. In fact, our results imply that the cooperative control problems can be exactly analyzed if all the associated network optimization problems are convex.

In this view, the present thesis has made a contribution towards a theory of cooperative control. In particular, this thesis provides a mean to identify problems in cooperative optimization and control that can be exactly analyzed in a structured way and proposes efficient tools for analyzing the behavior. The results of this thesis suggest that convexity and duality are the important themes for cooperative optimization and control.

6.2 Outlook

The area of cooperative optimization and cooperative control has already seen significant advances over the last years from both a practical as well as from a theoretical side. Several contributions towards a general theory have also been made in this thesis, where the concept of convexity turned out to be of major importance. From a theoretical perspective it is reasonable to distinguish between efficiently and non-efficiently solvable problems and to provide solution strategies for the first one. One can hope then, that the results obtained for the first problem class will lead at least to a better understanding of problems in the second class. In this direction, various relevant research questions remain open for future research.

An interesting open question is certainly, how the proposed approximation methods for distributed optimization can be used to handle non-convex optimization problems. In particular, many distributed model predictive control problems are faced with non-convexity. Non-convexity may result from a nonlinear system dynamics or directly from a non-convex group objective. For example, constraints in collision-avoidance for mobile robots lead to non-convex problem formulations. Although global optimality can certainly not be expected in non-convex problems, the polyhedral approximation and trajectory exchange idea still remains a promising approach. Trajectory exchange methods can easily provide feasible, although sub-optimal, solutions to non-convex problems. Additionally, the method we proposed in Chapter 3 has an interpretation as a distributed version of Dantzig-Wolfe decomposition. In centralized optimization theory, the Dantzig-Wolfe decomposition is one of the main tools for handling non-convex optimization problems, see for example the book Desaulniers et al. (2005). Therefore, from the present perspective it seems promising to investigate the use of polyhedral approximation methods for solving non-convex optimization problems in a distributed way.

In central non-convex robust optimization, much attention was dedicated over the last years to *sampling methods*. For an introduction to sampling methods for non-convex optimization we refer to the recent overview article of Garatti and Campi (2013) or the book Calafiore and Dabbene (2006). The appealing property of sampling methods is that they allow to quantify the quality of the final solution by the number of samples used to approximate the problem. Sampling methods are inherently related to the pessimizing oracle method we studied in Chapter 2.5. An open research question is now, how the results from sampling theory can be used in a distributed setup to quantify the sub-optimality of solutions computed by the cutting-plane consensus algorithm, when the algorithm is stopped premature.

Non-convexity is also the greatest obstacle for generalizing the results of the Chapters 4 and 5. Clearly, an open question is whether the duality results of Chapter 4 still hold if the node dynamics does not satisfy the equilibrium independent passivity property, or if the coupling nonlinearities are not monotonic. We have already seen on the Kuramoto oscillator that the problem can become significantly more difficult. However, several insights might still be obtained from the analysis. A deeper analysis of the duality relations in the non-convex case is certainly highly interesting. Another structural change of the cooperative control problem, breaking the passivity structure, is to consider a directed interaction topology. In many cooperative control problems, where a coupling between the dynamical systems is based on an information exchange, the underlying graph is directed. Various simple group objectives can then still be achieved, see e.g., Moreau (2005). However, the passivity framework studied in this thesis heavily depends on the undirected interaction structure, as the use of the incidence matrix to map edge states to node states preserved the passivity property in the closed-loop system. This was the basis for all duality results. Now, an open and interesting question is whether similar duality results as those presented in Chapter 4 can be established for networks with directed interaction structures.

To complete the discussion, we also want to point out that the control framework in Chapter 4 is intimately related to internal model control problems. In fact, the proposed framework can be understood as an internal model controller for systems driven by constant external signals. A natural question to ask now is whether the duality and optimality results established in this thesis still hold if the external signals are time-varying. A first attempt in this direction has been made in Bürger and De Persis (2013), without answering the question to the fullest extent. However, a general result relating internal model control under time-varying reference signals to a duality and network optimization framework promises interesting insights into the relation between control theory and Lagrange duality in time-varying problems.

Summarizing, this thesis has presented a theoretical framework for cooperative optimization and control. By restricting the discussion to certain important but well structured problem classes, it was possible to construct theoretical frameworks for both problems. The results derived in this thesis provide a fairly complete framework for understanding problems in the respective class. However, it remains an open challenge to extend the frameworks constructed here to broader problem classes.

Appendix A

Convex Analysis and Optimization Theory

The dominant property in modern optimization theory is convexity.[1] We review here some notions and basic concepts from convex analysis and convex optimization theory. The presentation follows the standard textbooks Rockafellar (1997) and Boyd and Vandenberghe (2004).

Convex Sets

Definition A.1 (Convex Set). *A set $\mathcal{C} \subset \mathbb{R}^q$ is said to be* convex *if the line segment connecting any two points in $\mathcal{C}$ is fully contained in $\mathcal{C}$. That is $x \in \mathcal{C}$ and $y \in \mathcal{C}$ implies $\lambda x + (1-\lambda)y \in \mathcal{C}$ for all $\lambda \in (0,1)$.*

We use the convention that the empty set $\mathcal{C} = \emptyset$ and the complete vector space, i.e., $\mathcal{C} = \mathbb{R}^q$ are convex sets. It follows directly from the definition that the intersection of an arbitrary collection of convex sets is convex. An important convex set if the *half-space* $\{x : a^\top x \leq b\}$, for some $a \in \mathbb{R}^q$ and $b \in \mathbb{R}$. A convex set is said to be *polyherdal* if it can be expressed as the intersection of a finite number of half-spaces.

A *convex combination* of a finite set of vectors $\{x_0, \ldots, x_m\}$ is a vector of the form $\lambda_0 x_0 + \cdots + \lambda_m x_m$, for some $\lambda_0 \geq 0, \ldots, \lambda_m \geq 0$, $\lambda_0 + \cdots + \lambda_m = 1$. The *convex hull* of a set $\mathcal{S} \subset \mathbb{R}^q$, denoted by conv$\mathcal{S}$, is the intersection of all convex sets containing $\mathcal{S}$. For any $\mathcal{S} \subset \mathbb{R}^n$, conv$\mathcal{S}$ consists of all the convex combinations of the elements of $\mathcal{S}$. The convex hull of a finite set is the set of all convex combinations. A set which is the convex hull of finitely many points $\{x_0, \ldots, x_m\}$ is called a *polytope*. If $\{x_0, \ldots, x_m\}$ is affinely independent, its convex hull is called an m-dimensional *simplex*. The points $x_0, \ldots, x_m$ are called the *vertices* of the simplex. Each point of the simplex can be uniquely expressed as a convex combination of the vertices, i.e.,

$$\Delta = \left\{ x \in \mathbb{R}^q \ : \ x = \sum_{i=0}^{m} \lambda_i x_i, \text{ for some } \sum_{i=0}^{m} \lambda_i = 1, \lambda_i \geq 0 \right\}.$$

In general, any point in a convex set $\mathcal{C} \subset \mathbb{R}^q$ lies in a m-dimensional simplex ($m \leq q$) with all vertices of the simplex being in $\mathcal{C}$. This result is more formally stated in Carathéodory's theorem.

[1] The statement of R. T. Rockafellar that the great watershed in optimization is between convexity and nonconvexity (Rockafellar, 1993) has nowadays been widely accepted.

Theorem A.2 (Carathéodory's Theorem). *Let $\mathcal{S}$ be any set of points and directions in $\mathbb{R}^q$, and let $\mathcal{C} = \operatorname{conv} \mathcal{S}$. Then $x \in \mathcal{C}$ if and only if x can be expressed as a convex combination of $n+1$ of the points and directions in $\mathcal{S}$ (not necessarily distinct).*

A point x in a set $\mathcal{S}$ is an *interior point*, if there exists an open set centered at x which is contained in $\mathcal{S}$. The set of all interior points is the *interior* of $\mathcal{S}$. In convex analysis the idea of an interior is often further refined. The *relative interior* of a set $\mathcal{S}$, denoted by relint $\mathcal{S}$, is its interior within the affine hull of $\mathcal{S}$. For a convex set $\mathcal{C}$ the relative interior can be expressed as

$$\operatorname{relint} \mathcal{C} := \{x \in \mathcal{C} \;:\; \text{for all } y \in \mathcal{C}, \text{there exists } \lambda > 1 : \lambda x + (1-\lambda) y \in \mathcal{C}\}.$$

Finally, we need the concept of separation between convex sets. Let $\mathcal{C}$ and $\mathcal{D}$ be two convex stets that do not intersect. We say that a hyperplane $h := \{x \in \mathbb{R}^q \;:\; a^\top x = b,\; a \neq 0\}$ *separates* $\mathcal{C}$ and $\mathcal{D}$ if $a^\top x \leq b$ for all $x \in \mathcal{C}$ and $a^\top x \geq b$ for all $x \in \mathcal{D}$. The hyperplane h separates $\mathcal{C}$ and $\mathcal{D}$ strongly if there exists $\epsilon > 0$ such that for all $x \in \mathcal{C}$ and all y satisfying $\|x - y\| \leq \epsilon$ it holds $a^\top y < b$, and for all $x \in \mathcal{D}$ and all y satisfying $\|x - y\| \leq \epsilon$ it holds $a^\top y > b$. The *separating hyperplane theorem* states that for any two non-empty disjoint convex sets a separating hyperplane exists. Forthermore, the following result is given in Rockafellar (1997).

Theorem A.3. *Let $\mathcal{C}$ and $\mathcal{D}$ be non-empty disjoint closed convex sets in $\mathbb{R}^q$. Then there exists a hyperplane separating $\mathcal{C}$ and $\mathcal{D}$ strongly.*

Convex and monotone functions The concept of convexity extends also to functions. Consider a function $\Phi : \mathcal{D} \subseteq \mathbb{R}^q \mapsto \mathbb{R}$ whose domain $\mathcal{D}$ is a convex set. The set

$$\operatorname{epi} \Phi := \{(x, \mu) \mid x \in \mathcal{D}, \mu \in \mathbb{R}, \mu \geq \Phi(x)\}$$

is called the *epigraph* of Φ. We say that Φ is a convex function on $\mathcal{D}$ if its epigraph is convex as a subset of $\mathbb{R}^{q+1}$. The *effective domain* of a convex function Φ on $\mathcal{D}$, denoted by dom Φ, is the projection on $\mathbb{R}^q$ of the epigraph of Φ, i.e.

$$\operatorname{dom} \Phi = \{x \in \mathcal{D} \;:\; \exists \mu, (x, \mu) \in \operatorname{epi} \Phi\} = \{x \in \mathcal{D} \;:\; \Phi(x) < +\infty\}.$$

A convex function is said to be *proper* if its epigraph is non-empty and contains no vertical lines.

Alternatively, often the following definitions for convex functions are employed:

Definition A.4. *A function $\Phi : \mathcal{D} \mapsto \mathbb{R}$ is said to be*

- convex *on a convex set $\mathcal{D}$ if for any two points $x, y \in \mathcal{D}$ and for all $\lambda \in [0,1]$,*

$$\Phi(\lambda x + (1-\lambda) y) \leq \lambda \Phi(x) + (1-\lambda) \Phi(y);$$

- strictly convex *if the inequality holds strictly whenever $x \neq y$; and*

- strongly convex *if there exists $\alpha > 0$ such that for any two points $x, y \in \mathcal{D}$, with $x \neq y$, and for all $\lambda \in [0,1]$*

$$\Phi(\lambda x + (1-\lambda) y) < \lambda \Phi(x) + (1-\lambda) \Phi(y) - \frac{1}{2} \lambda (1-\lambda) \alpha \|x - y\|^2.$$

A function is said to be *concave* on a convex set $\mathcal{D}$ if its negative is convex. An *affine* function on $\mathcal{D}$ is a function which is finite, convex and concave. A function $\Phi : \mathcal{D} \mapsto \mathbb{R}$ is convex if and only if

$$\Phi(\lambda_1 x_1 + \cdots + \lambda_m x_m) \leq \lambda_1 \Phi(x_1) + \ldots + \lambda_m \Phi(x_m)$$

whenever $\lambda_1 \geq 0, \ldots, \lambda_m \geq 0, \lambda_1 + \ldots + \lambda_m = 1$. This is known as *Jensen's Inequality*. A differentiable function Φ is convex if and only if its domain $\mathcal{D}$ is convex and

$$\Phi(y) \geq \Phi(x) + \nabla\Phi^\top(x)(y - x)$$

for all $y, x \in \mathcal{D}$. Furthermore, a continuous, twice differentiable function Φ is convex on a convex set $\mathcal{D}$ if and only if its Hessian matrix $\nabla^2\Phi(x)$ is positive definite for every $x \in \mathcal{D}$.

These conditions can be generalized to non-differentiable functions by considering the *subdifferential*. A vector g is said to be a *subgradient* of a function Φ at a point x if

$$\Phi(y) \geq \Phi(x) + g^\top(y - x), \quad \forall y \in \mathcal{D}.$$

If Φ is convex it has at least one subgradient at every point in relint$\mathcal{D}$. If Φ is additionally differentiable at x, $\nabla\Phi(x)$ is a subgradient of Φ at x. The set of all subgradients of Φ at x is called the *subdifferential* of Φ at x and is denoted by $\partial\Phi(x)$. The subdifferential is a closed convex set, and is nonempty if Φ is convex and finite near x. If the subdifferential contains only one point, i.e., $\partial\Phi(x) = \{g\}$, then Φ is differentiable and $g = \nabla\Phi(x)$. Let $\Phi_1, \ldots, \Phi_m$ be proper convex functions on $\mathbb{R}^q$, and let $\Phi = \Phi_1 + \cdots + \Phi_m$. Then

$$\partial\Phi(x) \subset \partial\Phi_1(x) + \cdots + \partial\Phi_m(x), \forall x \in \mathcal{D}.$$

The *subdifferential* of Φ is the multivalued mapping $\partial\Phi : x \mapsto \partial\Phi(x)$.

The subdifferential is closely related to an important duality concept. The *convex conjugate* of a convex function $\Phi(x)$ with convex domain $\mathcal{D}$, denoted $\Phi^\star$, is defined as (Rockafellar, 1997):

$$\Phi^\star(y) = \sup_{x \in \mathcal{D}}\{y^\top x - \Phi(x)\} = -\inf_{x \in \mathcal{D}}\{\Phi(x) - y^\top x\}.$$

If Φ is a proper convex function, then $x \in \partial\Phi^\star(y)$ if and only if $y \in \partial\Phi(x)$, see (Rockafellar, 1997, Cor. 23.5.1). Furthermore, $x \in \partial\Phi^\star(y)$ and $y \in \partial\Phi(x)$ if and only if $y^\top x = \Phi(x) + \Phi^\star(y)$.

Example A.5. *An important convex function is the* indicator function *for a convex set $\mathcal{D}$:*

$$I_\mathcal{D}(x) := \begin{cases} 0 & \text{if } x \in \mathcal{D} \\ +\infty & \text{if } x \notin \mathcal{D} \end{cases} \tag{A.1}$$

The subdifferential $\partial I_\mathcal{D}(x)$ is the set

$$\partial I_\mathcal{D}(x) = \{s \in \mathbb{R}^q \; : \; s^\top(y - x) \leq 0 \text{ for all } y \in \mathcal{D}\},$$

i.e., the normal cone of $\mathcal{C}$ at x. The convex conjugate is now

$$I_\mathcal{D}^\star(y) = \sup_{x \in \mathcal{D}} y^\top x.$$

Stronger statements about the structure of the subdifferential of a convex function and their connection to the convex conjugate can be established under a further smoothness assumption. A proper convex function Φ is said to be *essentially smooth* if it satisfies the following conditions on $\mathcal{C} = \text{int dom } \Phi$: (i) $\mathcal{C}$ is non-empty (ii) Φ is differentiable throughout $\mathcal{C}$ (iii) $\lim_{i\to\infty} |\nabla\Phi(x_i)| = +\infty$ whenever $x_1, x_2, \ldots,$ is a sequence in $\mathcal{C}$ converging to a boundary point x of $\mathcal{C}$. If Φ is a closed proper convex function, then $\partial\Phi$ is a single-valued mapping if and only if Φ is essentially smooth and in this case $\partial\Phi(x) = \{\nabla\Phi(x)\}$, see (Rockafellar, 1997, Theorem 26.1). A strong statement relating the gradients of a strictly convex function and its convex conjugate can now be made for essentially smooth functions, see (Rockafellar, 1997, Theorem 26.5).

Theorem A.6. *Let Φ be a essentially smooth, strictly convex function with $\mathcal{C} = \text{int dom } \Phi$ an open convex set. Then the gradient mapping $\nabla\Phi$ is one-to-one, continuous in both directions, and $\nabla\Phi^\star = (\nabla\Phi)^{-1}$.*

Finally, we want to connect convexity to another important function property, namely monotonicity.

Definition A.7. *A function $\phi : \mathbb{R}^q \mapsto \mathbb{R}^q$ is said to be*

- monotone *on a convex set $\mathcal{C}$ if $x \geq y$ implies $\phi(x) \geq \phi(y)$for all $x, y \in \mathcal{C}$;*
- strongly monotone *on $\mathcal{C}$ if there exists $\alpha > 0$ such that*

$$(\phi(x) - \phi(y))^\top (x - y) \geq \alpha \|x - y\|^2$$

for all $x, y \in \mathcal{C}$; and

- γ-co-coercive *on $\mathcal{C}$ if there exists $\gamma > 0$ such that*

$$(\phi(x) - \phi(y))^\top (x - y) \geq \gamma \|\phi(x) - \phi(y)\|^2$$

for all $x, y \in \mathcal{C}$ (see, e.g., Zhu and Marcotte (1995)).

Monotonicity and convexity are related concepts. *Kachurovskii's theorem* states that a continuously differentiable real function is convex if and only if its first derivative is monotone, and strictly convex if and only if its derivative is strongly monotone. Furthermore, the following result holds for scalar functions, see (Rockafellar, 1997, Theorem 24.2).

Theorem A.8. *Let $a \in \mathbb{R}$, and let ϕ be a monotone function from $\mathbb{R}$ to $[-\infty, +\infty]$ such that $\phi(a)$ is finite. Then the function Φ given by*

$$\Phi(x) = \int_a^x \phi(t)dt$$

is a well-defined closed proper convex function on $\mathbb{R}$.

We call the function Φ the integral of ϕ. Thus, the integral of a monotone function is convex, and, similarly, the integral of a strongly monotone, its integral Φ is strongly convex.

Convex Optimization & Lagrange Duality We turn our attention now to the problem of finding the "best" solution over a convex set. A *convex optimization problem* is the mathematical problem of finding the best solution of a convex objective function over a closed convex constraint set $\mathcal{X}$, i.e.,

$$\begin{aligned} \text{minimize} \quad & \Phi(x) \\ \text{subj. to} \quad & x \in \mathcal{X}. \end{aligned} \tag{A.2}$$

A convex optimization problem has the nice properties that a local optimum is always also a global optimum, and the set of global optima is convex. We denote in the following the optimal value of (A.2) with p^*. In general, one can assume that the objective function is *linear*. In fact, the minimization of an arbitrary convex function $\Phi(x)$ over a convex constraint set $\mathcal{X}$, can always be represented as a minimization of a linear function over a convex set of higher dimension by considering the *epigraph representation* of the problem

$$\begin{aligned} \text{minimize} \quad & \Phi(x) \\ \text{subj. to} \quad & x \in \mathcal{X} \end{aligned} \quad \Leftrightarrow \quad \begin{aligned} \text{minimize} \quad & \mu \\ \text{subj. to} \quad & (\mu, x) \in \tilde{\mathcal{X}}, \end{aligned} \tag{A.3}$$

where $\tilde{\mathcal{X}} = \{(\mu, x) \in \mathbb{R}^{q+1} : \mu \geq \Phi(x), x \in \mathcal{X}\}$ is a convex set.

Convex optimization problems can often (but not always) be expressed in the standard form

$$\begin{aligned} \text{minimize} \quad & \Phi_0(x) \\ \text{subj. to} \quad & \Phi_i(x) \leq 0, \ i \in \{1, \dots, n\} \\ & a_j^\top x = b_j, \ j \in \{1, \dots, m\}, \end{aligned} \tag{A.4}$$

where $\Phi_0, \dots, \Phi_n$ are convex functions. We define in the following $\mathcal{D} = \bigcap_{i=0}^n \operatorname{dom} \Phi_i$. The *Lagrangian function* associated to the convex problem in standard form (A.4) is defined as

$$\mathcal{L}(x, \lambda, \mu) := \Phi_0(x) + \sum_{i=1}^{n} \lambda_i \Phi_i(x) + \sum_{j=1}^{m} \mu_j (a_j^\top x - b_j), \tag{A.5}$$

where the variables λ_i and μ_j are called the *Lagrange multipliers*. The *Lagrange dual function* of (A.4) can now be derived as the minimum of the Lagrangian over the primal decision variable x, i.e.,

$$g(\lambda, \mu) := \inf_x \mathcal{L}(x, \lambda, \mu).$$

Note that if the Lagrangian function is unbounded in x, the dual function takes the value $-\infty$. Now, the *Lagrange dual problem* associated to (A.4) is the maximum over all Lagrange dual functions

$$\begin{aligned} \text{maximize} \quad & g(\lambda, \mu) \\ \text{subj. to} \quad & \lambda \geq 0. \end{aligned} \tag{A.6}$$

The optimal value of (A.6) is denoted with d^*. It always holds that $d^* \leq p^*$, which is known as *weak duality*. We say that *strong duality* holds if $d^* = p^*$.

Unfortunately, even for a convex problem of the form (A.4) strong duality does not hold in general. However, it holds under some additional assumptions on the constraints, the so-called constraint qualifications. One of the most important constraint qualifications is *Slater's condition: there exists* $x \in \text{relint}\mathcal{D}$ *such that* $\Phi_i(x) < 0, i \in \{1, \dots, n\}$ *and* $a_j^\top x = b_j$, $j \in \{1, \dots, m\}$. The point x is sometimes called a Slater point or a strictly feasible point. Slater's theorem states that strong duality holds if (A.4) is convex and a Slater point exists.

Lagrange duality has an important *saddle-point* interpretation. The dual problem (A.6) can be expressed as $\max_{\lambda \geq 0, \mu} \{\min_{x \in \mathcal{D}} \mathcal{L}(x, \lambda, \mu)\}$. Let now x^* be the optimal primal solution to (A.4), and λ^*, μ^* the optimal solution to (A.6). Then, the primal and dual solutions satisfy the saddle-point property

$$\mathcal{L}(x^*, \lambda, \mu) \leq \mathcal{L}(x^*, \lambda^*, \mu^*) \leq \mathcal{L}(x, \lambda^*, \mu^*)$$

for all $x \in \mathcal{D}$, $\lambda \in \mathbb{R}^n_{\geq 0}$ and $\mu \in \mathbb{R}^m$.

Finally, one can derive optimality conditions of (A.4) from the Lagrangian function. Assume strong duality holds and let x^* be a primal optimal solution and (λ^*, μ^*) be a dual optimal solution. Then

$$\begin{aligned} \nabla_x \mathcal{L}(x^*, \lambda^*, \mu^*) &= 0 \\ \Phi_i(x^*) &\leq 0, \quad i \in \{1, \dots, n\} \\ a_j^\top x^* &= b_j, \quad j \in \{1, \dots, m\} \\ \lambda_i^* &\geq 0, \quad i \in \{1, \dots, n\} \\ \lambda_i^* \Phi_i(x^*) &= 0, \quad i \in \{1, \dots, n\}. \end{aligned} \tag{A.7}$$

These conditions are known as the *Karush-Kuhn-Tucker* (KKT) conditions. If a convex optimization problem with differentiable objective and constraint functions satisfies Slater's condition, then the KKT conditions are necessary and sufficient for optimality. In fact, convexity and Slater's condition imply zero duality gap. Since the dual optimum is attained, x is optimal if and only if there are (λ, μ) that, together with x, satisfy the KKT conditions.

Appendix B

Dynamical Systems and Control Theory

We review here some basic concepts from dynamical systems, stability and feedback control. The presentation of this section follows the textbooks Khalil (2002) and Bullo et al. (2009). We consider here two different classes of dynamical systems: (i) discrete time continuous space and (ii) continuous time continuous space systems.

In general, a *control system* is a tuple $(\mathcal{X},\mathcal{U},\mathcal{X}_0,f)$, where $\mathcal{X}\subset\mathbb{R}^m$ is a non-empty set called the state-space, $\mathcal{U}\subset\mathbb{R}^p$ is a nonempty set containing the origin, called the input-value set, $\mathcal{X}_0\subset\mathcal{X}$ is the set of allowable initial conditions. In a *discrete time control system* the function $f:\mathbb{Z}_{\geq 0}\times\mathcal{X}\times\mathcal{U}\mapsto\mathcal{X}$ is called the evolution map. A function $x:\mathbb{Z}_{\geq 0}\mapsto\mathcal{X}$ satisfying for given initial condition $x(0)\in\mathcal{X}_0$ and input function $u:\mathbb{Z}_{\geq 0}\mapsto\mathcal{U}$ the difference equation

$$x(\ell+1)=f(\ell,x(\ell),u(\ell)),\quad \ell\in\mathbb{Z}_{\geq 0} \tag{B.1}$$

is called a *solution* of the discrete time control system In a *continuous time control system* the function $f:\mathbb{R}_{\geq 0}\times\mathcal{X}\times\mathcal{U}\mapsto T\mathcal{X}$, where $T\mathcal{X}$ is the tangent space of $\mathcal{X}$ is called the control vector field.[1] For a given initial condition $x(0)\in\mathcal{X}_0$ and control input function $u:\mathbb{R}_{\geq 0}\mapsto\mathcal{U}$, the map $x:\mathbb{R}_{\geq 0}\mapsto\mathcal{X}$ satisfying the ordinary differential equation

$$\dot{x}(t)=f(t,x(t),u(t)),\quad t\in\mathbb{R}_{\geq 0} \tag{B.2}$$

is called a solution of the continuous time control system. If f depends explicitly on the time ℓ or t, respectively, we say that the control system is *time-variant*, otherwise it is said to be *time-invariant.* We speak of a *dynamical system* instead of a control system if f does not depend on u.

A first important concept in control systems theory is the one of an *equilibrium point.* A point $x^*\in\mathcal{X}$ is an equilibrium point if the constant function $x(\ell)=x^*$ for all $\ell\in\mathbb{Z}_{\geq 0}$ ($x(t)=x^*$ for all $t\in\mathbb{R}_{\geq 0}$) is a solution of the dynamical system (B.1) ((B.2), respectively). It is clear, that for unforced time invariant systems discrete time control systems x^* is an equilibrium point if $x^*=f(x^*,0)$. Similarly, a point x^* is an equilibrium point of the *unforced* continuous time control system (B.2) if $0=f(x^*,0)$. Clearly, the idea of an equilibrium point can be directly generalized to systems with constant input signal $u(\ell)=u^*\in\mathcal{U},\ell\in\mathbb{Z}_{\geq 0}$ ($u(t)=u^*\in\mathcal{U},\ t\in\mathbb{R}_{\geq 0}$).

[1]Suppose $\mathcal{X}=\mathbb{R}^n$, then $f:\mathbb{R}_{\geq 0}\times\mathbb{R}^n\times\mathcal{U}\mapsto\mathbb{R}^n$.

We are now ready to introduce the notions of invariance, stability, and attractivity. For simplicity, we discuss first discrete time dynamical systems. Consider a time-invariant discrete time dynamical system with state space $\mathcal{X}$ and let $\mathcal{S} \subset \mathcal{X}$. The set $\mathcal{S}$ is *positively invariant* under the dynamical system if each solution $x(\ell)$ with initial condition in $\mathcal{S}$ remains in $\mathcal{S}$ for all times. A solution x *approaches* a set $\mathcal{S}$ if, for every neighborhood $\mathcal{W}$ of $\mathcal{S}$, there exists a time $\ell_0 > 0$ such that $x(\ell)$ takes values in $\mathcal{W}$ for all subsequent times $\ell \geq \ell_0$.

Definition B.1 (Stability). *Consider a time-invariant dynamical system. As set $\mathcal{S}$ is said to be*

1. stable *if, for any neighborhood $\mathcal{Y}$ of $\mathcal{S}$, there exists a neighborhood $\mathcal{W}$ of $\mathcal{S}$ such that every solution of* (B.1) *with initial condition in $\mathcal{W}$ remains in $\mathcal{Y}$ for all subsequent times;*

2. unstable *if it is not stable;*

3. locally attractive *if there exists a neighborhood $\mathcal{W}$ of $\mathcal{S}$ such that every evolution with initial condition in $\mathcal{W}$ approaches the set $\mathcal{S}$; and*

4. locally asymptotically stable *if it is stable and locally attractive.*

Furthermore, the set $\mathcal{S}$ is globally attractive *if every evolution of the dynamical system approaches it and it is* globally asymptotically stable *if it is stable and globally attractive.*

This definition holds clearly for both, discrete time and continuous time dynamical systems.

Given a discrete time dynamical system with state space $\mathcal{X}$ and evolution map f. Let $\mathcal{D} \subset \mathcal{X}$ be given. A function $V : \mathcal{X} \mapsto \mathbb{R}$ is non-increasing on $\mathcal{D}$ along the solutions $x(\ell)$ of the dynamical system if $V(f(x)) \leq V(x)$, for all $x \in \mathcal{D}$. A positive definite function, i.e., a function satisfying $V(0) = 0$ and $V(x) > 0$ for all $x \neq 0$, that is non-increasing along the solutions $x(\ell)$ is called a *Lyapunov function*. The following invariance result in presented in Bullo et al. (2009)[Theorem 1.19].

Theorem B.2. *Consider a time-invariant discrete time dynamical system with state space $\mathcal{X}$ and a continuous evolution map f. Let $\mathcal{S} \subset \mathcal{X}$ be a set that is a positively invariant with respect to the dynamical system and suppose all solutions starting in $\mathcal{S}$ being bounded. Suppose there exists a function $V : \mathcal{D} \mapsto \mathbb{R}$ that is continuous on $\mathcal{S}$ and non-increasing along the solutions of the dynamical system, i.e., $V(f(x)) \leq V(x)$, for all $x \in \mathcal{D}$. Then, all solutions starting in $\mathcal{S}$ approach a set of the form $V^{-1}(c) \cap \mathcal{M}$, where c is a real constant and $\mathcal{M}$ is the largest positively invariant set contained in $\{x \in \mathcal{S} \,:\, V(f(x)) = V(x)\}$.*

A similar result can be formalized now for continuous time control systems. We need to refine first the idea of non-decreasing functions for continuous time dynamical systems. Consider a continuous time dynamical system with state space $\mathcal{X}$ and control vector field f. For a function $V : \mathcal{X} \mapsto \mathbb{R}$, the *directional derivative* along the solutions of the dynamical system (B.2) is defined as

$$\dot{V} = \frac{\partial V}{\partial x} f(x). \tag{B.3}$$

In fact, given a specific solution $x(t)$ of the dynamical system $\dot{x} = f(x)$, the directional derivative is the time derivative along this solution, computed as $\dot{V} = \frac{d}{dt}V(x(t))$. A function $V(x)$ is *non-increasing* on a set $\mathcal{D} \subset \mathcal{X}$ along the solution of the continuous time dynamical system if $\dot{V}(x) \leq 0$ for all $x \in \mathcal{D}$. Now, we can state the invariance principle for continuous time systems, known as *LaSalle's theorem* (see Khalil (2002)[Theorem 4.4], Bullo et al. (2009)[Theorem 1.20]).

Theorem B.3. *Consider a time-invariant continuous time dynamical system with state space $\mathcal{X}$ and control vector field f. Suppose f is continuously differentiable. Let $\mathcal{S} \subset \mathcal{X}$ be a set that is positively invariant with respect to the dynamical system and suppose all solutions starting in $\mathcal{S}$ being bounded. Suppose there exists a function $V : \mathcal{D} \mapsto \mathbb{R}$ that is continuous on $\mathcal{S}$ and non-increasing along the solutions of the dynamical system, i.e., $\frac{\partial V}{\partial x} f(x) \leq 0$ for all $x \in \mathcal{S}$. Then, all solutions starting in $\mathcal{S}$ approach a set of the form $V^{-1}(c) \cap \mathcal{M}$, where c is a real constant and $\mathcal{M}$ is the largest positively invariant set contained in $\{x \in \mathcal{S} \,:\, \dot{V}(x) = 0\}$.*

Sometimes the assumptions on imposed by LaSalle's theorem are too restrictive and other invariance-like results are required. The following result is known as Barbalat's lemma (Khalil, 2002)[Lemma 8.2].

Lemma B.4. *Let $\phi : \mathbb{R} \mapsto \mathbb{R}$ be a uniformly continuous function on $[0, \infty)$. Suppose that $\lim_{t\to\infty} \int_0^t \phi(\tau) d\tau$ exists and is finite. Then $\phi(t) \to 0 \quad as \quad t \to \infty$.*

Barbalat's lemma is useful for analyzing the aymptotic behavior of continuous time dynamical control systems.

Theorem B.5. *Consider a time-invariant continuous time dynamical system with state space $\mathcal{X}$ and control vector field f, with $f(x)$ locally Lipschitz in x Suppose there exists a continuously differentiable function $V : \mathcal{X} \mapsto \mathbb{R}$ such that*

$$\frac{\partial V}{\partial x} f(x) \leq -W(x), \; \forall x \in \mathcal{X}, \tag{B.4}$$

where $W(x)$ is a continuous positive semidefinite function on $\mathcal{X}$. Then, all solutions $x(t)$ are bounded and satisfy

$$W(x(t)) \to 0 \quad as \quad t \to \infty.$$

We turn our attention now to control systems with input and outputs. We restrict the discussion in the following only to continuous time control systems, as the following concepts are in this thesis only important for this system class. A continuous time control system with input and outputs takes the form

$$\dot{x} = f(x, u) \; y = h(x, u), \tag{B.5}$$

where $x \in \mathcal{X} \subset \mathbb{R}^n$ and $u \in \mathcal{U} \subset \mathbb{R}^p$ and $y \in \mathcal{Y} \subset \mathbb{R}^p$. We restrict our attention here to *square* control systems with the same number of inputs and outputs. A key role in this thesis takes the system theoretic concept *passivity*. We assume in the following that $x^* = 0$ is an equilibrium point of the unforced control system (B.5), that is $f(0, 0) = 0$. In the following, if we refer to a positive definite function then this function is assumed to be positive definite with respect to this equilibrium point. Additionally, given a function $S(x)$ we use the short-hand notation $\dot{S} := \frac{\partial S}{\partial x} f(x, u)$.

Definition B.6. *The system* (B.5) *is said to be* passive *if there exists a positive semidefinite function $S(x)$ (called storage function) such that $y^\top u \geq \dot{S}$. Furthermore, it is said to be* output strictly passive *if $y^\top u \geq \dot{S} + y^\top \phi(y)$ and $y^\top \phi(y) > 0$ for $y \neq 0$; and* strictly passive *if $y^\top u \geq \dot{S} + \psi(x)$ for some positive definite function ψ.*

Roughly speaking, passivity generalizes the idea of energy conservation to general control systems of the form (B.5). An often employed interpretation relates to electric circuits: if u is taken as voltage and y is taken as current, then $y^\top u$ is the power flow into the network, see Khalil (2002). While passivity might be hard to verify for general nonlinear control systems, elegant conditions exist for linear dynamical systems of the form

$$\begin{aligned} \dot{x}(t) &= Ax(t) + Bu(t) \\ y(t) &= Cx(t) + Dx(t). \end{aligned} \tag{B.6}$$

The following result, known as *Kalman-Yakubovich-Popov (KYP) lemma*, establishes sufficient conditions for passivity of linear systems.

Lemma B.7. *Consider the linear time-invariant system* (B.6)*, where (A, B) is controllable and (A, C) is observable. Then,* (B.6) *is strictly passive if there exist matrices $P = P^\top > 0$, L, W, and a positive constant ϵ such that*

$$\begin{aligned} PA + A^\top P &= -L^\top L - \epsilon P \\ PB &= C^\top - L^\top W \\ W^\top W &= D + D^\top. \end{aligned} \tag{B.7}$$

It is passive, if (B.7) *holds with $\epsilon = 0$.*

An extension of this result is known for nonlinear input affine systems of the form

$$\begin{aligned} \dot{x}(t) &= f(x(t)) + G(x(t))u(t) \\ y(t) &= h(x), \end{aligned} \tag{B.8}$$

where f is locally Lipschitz, G and h are continuous, $f(0) = 0$, and $h(0) = 0$.

Lemma B.8. *Consider the system* (B.8)*. The system is output strictly passive if there exists a positive semidefinite storage function $S(x)$ and some positive constant ϵ such that*

$$\begin{aligned} &\frac{\partial S}{\partial x} f(x) \leq -\epsilon h^\top(x) h(x) \\ &\frac{\partial S}{\partial x} G(x) = h^\top(x). \end{aligned} \tag{B.9}$$

It is passive if (B.9) *holds for $\epsilon = 0$.*

Appendix C

Graph Theory

We introduce here some basic notions from graph theory. The presentation of this section follows the books Godsil and Royle (2001) and Bullo et al. (2009).

A *directed graph* (also called *digraph*) is the pair $\mathcal{G} = (\mathbf{V}, \mathbf{E})$, where $\mathbf{V} = \{v_1, \ldots, v_n\}$ represents a set of n *nodes* (or *vertices*) and $\mathbf{E} = \{e_1, \ldots, e_m\}$ is a set of *edges*. Each edge e_k represents an ordered pair of vertices (v_i, v_j), where v_i is the initial node of edge e_k and v_j is the terminal node of edge e_k. We will sometimes write $(v_i, v_j) \in \mathbf{E}$. In an *undirected graph* (also called *graph*) the edges e_k represent an unordered pair of vertices $\{v_i, v_j\}$ and no distinction is made whether the edge originates or terminates at a node. It is always possible to associate to a digraph a graph, by simply ignoring the direction on the edges. Similarly, one can always derive a digraph from a graph by introducing an arbitrary direction on the edges.

Remark C.1. *The notation we introduce here assigns to each node and each edge a unique identifier, i.e.,* $\mathbf{V} = \{v_1, \ldots, v_n\}$ *and* $\mathbf{E} = \{e_1, \ldots, e_m\}$. *Although this notation is precise, we use in this thesis sometimes a simplified notation and write simply* $i \in \mathbf{V}$ *or* $k \in \mathbf{V}$. *Strictly speaking, this notation does not make sense since* $i \in \mathbb{N}$ *cannot refer to both, a node or an edge. However, in most cases this notation can be used unambiguously.*

In a digraph $\mathcal{G}$ with an edge $e_k = (v_i, v_j) \in \mathbf{E}$, the node v_i is called the *in-neighbor* of node v_j, and node v_j is called the *out-neighbor* of v_i. We also say than that the two nodes v_i and v_j are *adjacent*, and that the nodes v_i and v_j are *incident* to the edge e_k. The set of all in-neighbors and out-neighbors of node v_i are denoted by $\mathcal{N}_I(v_i)$ and $\mathcal{N}_O(v_i)$, respectively. The *in-degree* and the *out-degree* are the cardinality of the sets $\mathcal{N}_I(v_i)$ and $\mathcal{N}_O(v_i)$, respectively. In an undirected graph $\mathcal{G}$ the nodes v_i and v_j are said to be *neighbors* if they are connected by an edge $e_k \in \mathbf{E}$. The set of all neighbors of node v_i is denoted by $\mathcal{N}(v_i)$. The degree of v_i is the cardinality of $\mathcal{N}(v_i)$.

We review next some *connectivity* notions. A *path* in a graph is an ordered sequence of nodes such that any pair of consecutive vertices in the sequence is an edge of the graph. A graph is *connected* if there exists a path between any two vertices. If a graph is not connected, then it is composed of multiple *connected components*. A path is *simple* if no nodes appear more than once in it, except possibly for the initial and final vertex. A *cycle* is a simple path that starts and ends at the same node. A *Hamiltonian path* is a path that meets every node, and a *Hamiltonian cycle* is a cycle that meets every node. A connected acyclic graph is a *tree*. A graph is a tree, if and only if it is connected and has exactly $n-1$ edges. A *forest* is a graph that can be expressed as the disjoint union of trees.

The connectivity notion needs to be more refined for digraphs. A *directed path* in a digraph is an ordered sequence of nodes such that any ordered pair of nodes appearing consecutively in the sequence is an edge of the digraph. A *cycle* in a digraph is a directed path that starts and ends at the same node and that contains no repeated nodes except for the initial and final node. A digraph is *acyclic* if it contains no cycles. A node of a digraph is *globally reachable* if it can be reached from any other vertex by traversing a directed path. A digraph is *strongly connected* if every vertex is globally reachable.

Given two nodes v_i and v_j of a digraph $\mathcal{G}$, the *distance* from v_i to v_j, denoted by $\text{dist}(v_i, v_j)$ is the smallest length of a directed path from v_i to v_j, or $+\infty$ if there is no directed path from v_i to v_j. The *diameter* of a digraph is

$$\text{diam}(\mathcal{G}) = \max\{\text{dist}(v_i, v_j) | v_i, v_j \in \mathbf{V}\}.$$

Some special and important (undirected) graphs are the following:

- A *path graph* with n nodes is a graph, whose nodes can be ordered as $\{v_1, \ldots, v_n\}$ such that the edge set consists exactly of the $n-1$ edges $\{v_{i-1}, v_i\}$ for $i \in \{2, \ldots, n\}$, i.e., it is a graph that consists of a single path. A path graph has diameter $n-1$.
- A *cycle graph* with n nodes is a graph that consists of a single cycle. It has exactly n edges and a diameter of $\left\lfloor \frac{n}{2} \right\rfloor$.
- *Circulant graphs* are graphs for which there exists an ordering of the node set as $\{v_1, \ldots, v_n\}$, such that if two nodes v_i and v_j are adjacent, then every two nodes v_r and v_s with $s = (l - i + j) \bmod n$ are adjacent.
- A *regular graph* is one where all nodes have the same number of neighbors. A graph is called a k-regular graph is all nodes have exactly k neighbors.
- In a *complete graph* each pair of is connected by an edge. The complete graph has all possible $\frac{n(n-1)}{2}$ edges and has a diameter of 1.
- Finally, *Erdös-Rényi graphs* are random graphs with a fixed node set $|\mathbf{V}| = n$. An Erdös-Rényi graph is constructed by connecting the nodes randomly, i.e., an edge is included with probability p independent from every other edge. Almost any Erdös-Rényi graph with $p = (1+\epsilon)\frac{\ln n}{n}$ for some positive constant ϵ is almost surely connected, see Erdös and Rényi (1960). Additionally, random graphs usually have a small diameter. It has been shown that the range of values in which the diameters can be is very small and centered around $d = \frac{\ln(n)}{\ln(np)}$, see Albert and Barabási (2002).

Besides the abstract definition, graphs can also be represented by matrices. This leads to the mathematical area of *algebraic graph theory.* Several important matrices can be defined for graphs and digraphs. The *adjacency matrix* $A \in \mathbb{R}^{|\mathbf{V}|\times|\mathbf{V}|}$ of a digraph is the integer matrix with rows and columns indexed by the elements of $\mathbf{V}$, such that the i, j-th entry is equal to the number of edges connecting v_i and v_j (in a directed sense). That is $[A]_{ij} = 1$ if v_i is an out-neighbor of v_j and zero otherwise. If the graph is directed, A is a symmetric matrix. Let D_{in} be a diagonal matrix with the i-th diagonal elements being the in-degree of node i, i.e., $[D_{in}]_{ii} = |\mathcal{N}_I(v_i)|$. The *Laplacian matrix* of $\mathcal{G}$ is defined as $L = D_{in} - A$. The *incidence matrix* E of a digraph $\mathcal{G}$ is a $\{0, \pm 1\}$ matrix, with rows indexed by the elements of $\mathbf{V}$ and columns indexed by elements of $\mathbf{V}$ and $[E]_{ik} = +1$ if i is the initial node of the edge k, $[E]_{ik} = -1$ if i is the terminal node of the edge k, and $[E]_{ik} = 0$ otherwise. Now, the symmetric matrix $L = EE^\top$ is the Laplacian matrix of the undirected graph, resulting from $\mathcal{G}$ by removing the direction of the edges.

Bibliography

D. Aeyels and F. De Smet. A mathematical model for the dynamics of clustering. *Physica D: Nonlinear Phenomena*, 237(19):2517–2530, 2008.

D. Aeyels and F. De Smet. Cluster formation in a time-varying multi-agent system. *Automatica*, 47(11):2481–2487, 2011.

R. Albert and A.-L. Barabási. Statistical mechanics of complex networks. *Reviews of Modern Physics*, 74(1):47–97, 2002.

A. B. Amenta. *Helly Theorems and Generalized Linear Programming.* PhD thesis, University of California at Berkeley, 1993.

M. Arcak. Passivity as a design tool for group coordination. *IEEE Transactions on Automatic Control*, 52(8):1380–1390, 2007.

J. Bachrach and C. Taylor. *Handbook of sensor networks*, chapter Localization in sensor networks, pages 277–310. John Wiley and Sons, Inc., 2005.

H. Bai, M. Arcak, and J. Wen. *Cooperative control design: A systematic, passivity-based approach.* Springer, New York, NY, 2011.

M. Bando, K. Hasebe, A. Nakayama, A. Shibata, and Y. Sugiyama. Dynamical model of traffic congestion and numerical simulation. *Physical Review E*, 51(2):1035–1042, 1995.

D. Bauso, F. Blanchini, and R. Pesenti. Robust control strategies for multi inventory systems with average flow constraints. *Automatica*, 42(8):1255 – 1266, 2006.

S. Behnke. Kooperierende mobile Roboter. *at - Automatisierungstechnik*, 61(4):233 – 244, 2013.

A. Ben-Tal and A. Nemirovski. Robust solutions of uncertain linear programs. *Operations Research Letters*, 25:1–13, 1999.

A. Ben-Tal, L. E. Ghaoui, and A. Nemirovski. *Robust Optimization.* Princeton University Press, 2009.

D. Bertsekas. *Network Optimization: Continuous and Discrete Models.* Athena Scientific, Belmont, Massachusetts, 1998.

D. Bertsekas. Min Common / Max Crossing Duality: A Geometric View of Conjugacy in Convex Optimization. Technical Report Report LIDS-P-2796, Lab. for Information and Decision Systems, MIT, 2009.

D. P. Bertsekas and J. N. Tsitsiklis. *Parallel and Distributed Computations: Numerical Methods.* Athena Scientific, Belmont, Massachusetts, 1997.

P. Biswas and Y. Ye. A distributed method for solving semidefinite programs arising from ad hoc wireless sensor network localization. *Multiscale Optimization Methods and Applications*, 82:64–84, 2006.

V. Blondel, J. M. Hendrickx, and J. N. Tsitsiklis. On Krauses multi-agent consensus model with state-dependent connectivity. *IEEE Transactions on Automatic Control*, 54: 2586–2597, 2009.

E. Bompard, D. Wu, and F. Xue. The concept of betweenness in the analysis of power grid vulnerability. In *Complexity in Engineering, (COMPENG)*, pages 52–54, Rome, 2010.

S. Boyd and L. Vandenberghe. *Convex Optimization*. Cambridge University Press, 2004.

S. Boyd, N. Parikh, E. Chu, B. Peleato, and J. Eckstein. Distributed optimization and statistical learning via alternating direction method of multipliers. *Foundations and Trends in Machine Learning*, 3(1):1 – 122, 2010.

L. Bregman. The relaxation method of finding the common point of convex sets and its application to the solution of problems in convex programming. *USSR Computational Mathematics and Mathematical Physics*, 7(3):200 – 217, 1967.

F. Bullo, J. Cortés, and S. Martínez. *Distributed Control of Robotic Networks*. Applied Mathematics Series. Princeton University Press, 2009.

M. Bürger and C. De Persis. Internal models for nonlinear output agreement and optimal flow control. In *IFAC Symposium on Nonlinear Control Systems (NOLCOS)*, 2013. accepted.

M. Bürger, G. S. Schmidt, and F. Allgöwer. Preference based group agreement in cooperative control. In *Proc. of 8th IFAC Symposium on Nonlinear Control Systems (NOLCOS)*, pages 149–154, Bologna, Italy, 2010.

M. Bürger, G. Notarstefano, and F. Allgöwer. Locally constrained decision making via two-stage distributed simplex. In *Proc. IEEE Conference on Decision and Control, European Control Conference*, pages 5911 – 5916, Orlando, FL., Dec. 2011a.

M. Bürger, G. Notarstefano, F. Allgöwer, and F. Bullo. A distributed simplex algorithm and the multi-agent assignment problem. In *Proc. of American Control Conference*, pages 2639–2644, San Francisco, CA, USA, 2011b.

M. Bürger, D. Zelazo, and F. Allgöwer. Network clustering: A dynamical systems and saddle-point perspective. In *Proc. of IEEE Conference on Decision and Control*, pages 7825–7830, Orlando, FL, USA, Dec. 2011c.

M. Bürger, G. Notarstefano, and F. Allgöwer. Distributed robust optimization via cutting-plane consensus. In *Proc. IEEE Conference on Decision and Control*, pages 7457–7463, Maui, HI, USA, Dec. 2012a.

M. Bürger, G. Notarstefano, F. Bullo, and F. Allgöwer. A distributed simplex algorithm for degenerate linear programs and multi-agent assignments. *Automatica*, 48(9):2298–2304, July 2012b.

M. Bürger, D. Zelazo, and F. Allgöwer. Combinatorial insights and robustness analysis for clustering in dynamical networks. In *Proc. of the American Control Conference*, pages 454–459, Montreal, Canada, 2012c.

M. Bürger, G. Notarstefano, and F. Allgöwer. From non-cooperative to cooperative distributed MPC: A simplicial approximation perspective. In *Proc. of the European Control Conference*, pages 2795–2800, Zürich, CH, July 2013a.

M. Bürger, D. Zelazo, and F. Allgöwer. Hierarchical clustering of dynamical networks using a saddle-point analysis. *IEEE Transactions on Automatic Control*, 58(1):113–124, 2013b.

M. Bürger, D. Zelazo, and F. Allgöwer. Duality and network theory in passivity-based cooperative control. *Automatica*, 2013c. submitted Jan. 2013.

M. Bürger, D. Zelazo, and F. Allgöwer. On the steady-state inverse-optimality of passivity-based cooperative control. In *4th IFAC Workshop on Distributed Estimation and Control*, pages 138–143, Koblenz, Germany, 2013d.

M. Bürger, G. Notarstefano, and F. Allgöwer. A polyhedral approximation framework for convex and robust distributed optimization. *IEEE Transactions on Automatic Control*, February 2014.

G. Calafiore. Random Convex Programs. *SIAM Journal on Optimization*, 20(6):3427–3464, 2010.

G. Calafiore and F. Dabbene, editors. *Probabilistic and Randomized Methods for Desing under Uncertainty*. Springer, 2006.

L. Carlone, V. Srivastava, F. Bullo, and G. C. Calafiore. Distributed random convex programming via constraints consensus. *SIAM Journal of Control and Optimization*, July 2012. Submitted.

N. Chopra and M. Spong. *Advances in Robot Control, From Everyday Physics to Human-Like Movements*, chapter Passivity-based Control of Multi-Agent Systems, pages 107–134. Springer, 2006.

B. Curtis Eaves and W. I. Zangwill. Generalized cutting plane algorithms. *SIAM Journal of Control and Optimization*, 9(4):529–542, 1971.

G. Dantzig. Linear programming under uncertainty. *Management Science*, 1(3-4):197–206, 1955.

G. Dantzig and P. Wolfe. The decomposition algorithm for linear programs. *Econometrica*, 29(4):767–778, 1961.

G. B. Dantzig. *Linear Programming and Extensions*. Princeton University Press, 1963.

C. De Persis. Balancing time-varying demand-supply in distribution networks: an internal model approach. In *Proc. of European Control Conference*, 2013. to appear.

C. De Persis and B. Jayawardhana. Coordination of passive systems under quantized measurements. *SIAM Journal on Control and Optimization*, 50(6):3155 – 3177, 2012.

F. De Smet and D. Aeyels. Clustering in a network of non-identical and mutually interacting agents. *Proceedings of the Royal Society A*, 465:745–768, 2009.

P. DeLellis, M. Di Bernardo, and G. Russo. On QUAD, lipschitz, and contracting vector fields for consensus and synchronization of networks. *IEEE Transactions On Circuits and Systems - I*, 58(3):573–576, 2011.

G. Desaulniers, J. Desrosiers, and M. Solomon, editors. *Column Generation.* Springer, 2005.

M. D. Doan, T. Keviczky, and B. D. Schutter. A distributed optimization-based approach for hierarchical MPC of large scale systems with coupled dynamics and constraints. In *Proc. of the IEEE Conf. on Decision and Control and European Control Conference*, pages 5236–5241, 2011.

L. Doherty, K. S. J. Pister, and L. E. Ghaoui. Convex position estimation in wireless sensor networks. In *Proc. of the IEEE Conference on Computer Communications Societies*, volume 3, pages 1655–1663, 2001.

J. Dongarra and F. Sullivan. Guest Editors' introduction: The top 10 algorithms. *Computing in Science and Engineering*, 2(1):22–23, 2000.

F. Dörfler and F. Bullo. Exploring synchronization in complex oscillator networks. In *Proceedings of the IEEE Conf. on Decision and Control*, pages 7157–7170, Maui, HI, 2012.

J. R. Dunham, D. G. Kelly, and J. W. Tolle. Some experimental results concerning the expected number of pivots for solving randomly generated linear programs. Technical Report 77-16, Operations Research and System Analysis Department, University of North Carolina and Chapel Hill, 1977.

M. Dyer, N. Megiddo, and E. Welzl. *Handbook of Discrete and Computational Geometry*, chapter Linear Programming, pages 999–1014. Chapman and Hall, 2004.

P. Erdös and A. Rényi. On the evolution of random graphs. *Publications of the Mathematical Institute of the Hungarian Academy of Sciences*, 5:17–61, 1960.

M. Escalona-Morán, M. G. Cosenza, P. Guillén, and P. Coutin. Synchronization and clustering in electroencephalographic signals. *Chaos, Solitons & Fractals*, 31(4):820–825, 2007.

C. Ford and D. Fulkerson. Maximal Flow Through a Network. *Canadian Journal of Mathematics*, 8:399–404, 1956.

S. Fortunato. Community detection in graphs. *Physics Reports*, 486:75–174, 2010.

A. Franchi, P. Robuffo Giordano, C. Secchi, H. I. Son, and H. H. Bülthoff. Passivity-based decentralized approach for the bilateral teleoperation of a group of UAVs with switching topology. In *Proc. of IEEE International Conference on Robotics and Automation*, pages 898–905, Piscataway, NJ, USA, 2011.

B. A. Francis. The linear multivariable regulator problem. *SIAM Journal on Control and Optimization*, 14:486–505, 1976.

S. Garatti and M. C. Campi. Modulating robustness in control design: Principles and algorithms. *IEEE Control Systems Magazine*, 33(2):36–51, 2013.

B. Gärtner and E. Welzl. Linear programming – randomization and abstract frameworks. In *Proc. of the Symp. on Theoretical Aspects of Computer Science*, pages 669–687, 1996a.

B. Gärtner and E. Welzl. *Linear Programming-Randomization and Abstract Frameworks*, volume 1046 of *Lecture Notes in Computer Science*, chapter Symposium on Theoretical Aspects of Computer Science, pages 669–687. Springer, 1996b.

P. Giselsson and A. Rantzer. Distributed model predictive control with suboptimality estimates. In *Proc. of the IEEE Conf. on Decision and Control*, pages 7272–7277, 2010.

C. Godsil and G. Royle. *Algebraic Graph Theory*. Springer, 2001.

S. K. Goyal and B. Giri. Recent trends in modeling of deteriorating inventory. *European Journal of Operational Research*, 134:1–16, 2001.

L. Grüne and K. Worthmann. *Distributed Decision Making and Control*, chapter A Distributed NMPC Scheme without Stabilizing Terminal Constraints, pages 261–287. Lecture Notes in Control and Information Sciences. Springer, 2012.

J. K. Hale. Diffusive coupling, dissipation, and synchronization. *Journal of Dynamics and Differential Equations*, 9:865/9/1–1, 1997.

D. Helbing and B. Tilch. Generalized force model of traffic dynamics. *Physical Reviews E*, 58(1):133–138, July 1998.

J. M. Hendrickx, A. Olshevsky, and J. Tsitsiklis. Distributed anonymous discrete function computation. *IEEE Transactions on Automatic Control*, 56(10):2276–2289, 2011.

G. H. Hines, M. Arcak, and A. K. Packard. Equilibrium-independent passivity: A new definition and numerical certification. *Automatica*, 47(9):1949–1956, 2011.

A. Isidori and C. Byrnes. Output regulation of nonlinear systems. *IEEE Transactions on Automatic Control*, 35(2):131 –140, 1990.

A. Jadbabaie, J. Lin, and A. S. Morse. Coordination of groups of mobile agents using nearest neighbor rules. *IEEE Transactions on Automatic Control*, 48(6):988–1001, 2003.

B. Jayawardhana, R. Ortega, E. Gracia-Canseco, and F. Castanos. Passivity on nonlinear incremental systems: Application to PI stabilization of nonlinear RLC circuits. *Systems and Control Letters*, 56:618–622, 2007.

B. Johansson, M. Rabi, and M. Johansson. A randomized incremental subgradient method for distributed optimizaton in networked systems. *SIAM Journal on Optimization*, 20(3): 1157–1170, 2009.

C. N. Jones, E. C. Kerrigan, and J. M. Maciejowski. Lexicographic perturbation for multiparametric linear programming with applications to control. *Automatica*, 43(10): 1808–1816, 2007.

J. E. Kelley. The cutting plane method for solving convex programs. *SIAM Journal on Applied Mathematics*, 8:703–712, 1960.

H. Khalil. *Nonlinear Systems.* Prentice Hall, Upper Saddle River, New Jersey, 2002.

H. Konno, N. Kawadai, and H. Tuy. Cutting-plane algorithms for nonlinear semi-definite programming problems with applications. *Journal of Global Optimization*, 25:141–155, 2003.

M. Kraning, E. Chu, J. Lavaei, and S. Boyd. Message passing for dynamic network energy management. Technical report, Stanford University, 2012.

K. Krishnan and J. Mitchell. A unifying framework for several cutting plane methods for semidefinite programming. *Optimization Methods and Software*, 21:57–74, 2006.

L. S. Lasdon. *Optimization Theory for Large Scale Systems.* Courier Dover Publications, 2002.

D. Lazer, A. Pentland, L. Adamic, S. Aral, and A.-L. Barabasi. Computational Social Science. *Science*, 323(5915):721–723, 2009.

B. Lesieutre, S. Roy, and A. Pinar. Power system extreme event screening using graph partitioning. In *Proc. of the North American Power Symposium*, pages 503–510, 2006.

T. Liu, J. Zhao, and D. Hill. Incremental-dissipativity-based synchronization of interconnected systems. In *Proceedings of the 18th IFAC World Congress*, Milano, Italy, 2011.

M. Lopez and G. Still. Semi-infinite programming. *European Journal of Operational Research*, 180:491–518, 2007.

M. Lorenzen. Distributed robust power grid management via adjustable robust optimization. Master's thesis, Institute for Systems Theory and Automatic Control, University of Stuttgart, 2013.

M. Lorenzen, M. Bürger, G. Notarstefano, and F. Allgöwer. Robust economic dispatch problem and a distributed solution scheme. In *4th IFAC Workshop on Distributed Estimation and Control in Networked Systems*, pages 75–80, Koblenz, Germany, 2013.

S. H. Low and D. E. Lapsley. Optimization flow control - i: Basic algorithm and convergence. *IEEE/ACM Transactions on Networking*, 7(6):861–874, 1999.

S. H. Low, F. Paganini, and J. C. Doyle. Internet congestion control. *IEEE Control Systems Magazine*, 22(1):28 – 43, 2002.

D. G. Luenberger. *Introduction to Linear and Nonlinear Programming.* Addison-Wesley Publishing Company, 1973.

Z. Ma, Z. Liu, and G. Zhang. A new method to realize cluster synchronization in connected chaotic networks. *Chaos*, 16:23103, 2006.

O. L. Mangasarian. Least-norm linear programming solution as an unconstrained optimization problem. *Journal of Mathematical Analysis and Applications*, 92:240–251, 1983.

O. L. Mangasarian and R. R. Meyer. Nonlinear perturbation of linear programs. *SIAM Journal on Optimization*, 17(6):745–752, 1979.

M. Mesbahi and M. Egerstedt. *Graph Theoretic Methods in Multiagent Networks*. Princeton University Press, 2010.

L. Moreau. Stability of multiagent systems with time-dependent communication links. *IEEE Transactions on Automatic Control*, 50(2):169 – 182, 2005.

M. A. Müller, M. Reble, and F. Allgöwer. Cooperative control of dynamically decoupled systems via distributed model predictive control. *International Journal of Robust and Nonlinear Control*, 22(12):1376–1397, 2012.

K. G. Murty. *Linear Programming*. Wiley, 1983.

A. Mutapcic and S. Boyd. Cutting-set methods for robust convex optimization with pessimizing oracles. *Optimization Methods and Software*, 24:381–406, 2009.

I. Necoara, V. Nedelcu, and I. Dumitrache. Parallel and distributed optimization methods for estimation and control in networks. *Journal of Process Control*, 21:756–766, 2011.

A. Nedic and A. Ozdaglar. Distributed subgradient methods for multi-agent optimization. *IEEE Transactions on Automatic Control*, 54(1):48–61, 2009.

A. Nedic, A. Ozdaglar, and P. A. Parrilo. Constrained consensus and optimization in multi-agent networks. *IEEE Transactions on Automatic Control*, 55(4):922–938, 2010.

G. Notarstefano and F. Bullo. Network abstract linear programming with application to minimum-time formation control. In *Proc of the IEEE Conference on Decision and Control*, pages 927–932, New Orleans, USA, 2007.

G. Notarstefano and F. Bullo. Distributed abstract optimization via constraints consensus: Theory and applications. *IEEE Transactions on Automatic Control*, 56(10):2247–2261, 2011.

R. Olfati-Saber, J. A. Fax, and R. M. Murray. Consensus and cooperation in networked multi-agent systems. *Proceedings of the IEEE*, 95(1):215–233, 2007.

A. Olshevsky and J. Tsistiklis. Convergence speed in distributed consensus and averaging. *SIAM Journal of Control and Optimization*, 48(1):33–55, 2009.

J. Pannek. Parallelizing a state exchange strategy for noncooperative distributed NMPC. *Systems and Control Letters*, 62:29–36, 2013.

A. Pinar, Y. Fogel, and B. Lesieutre. The inhibiting bisection problem. Technical report, Lawrence Berkeley National Laboratory, 2006.

Power Systems Test Case Archive. http://ee.washington.edu/research/pstca/, 2013.

R. Reemtsen. Some outer approximation methods for semi-infinite optimization problems. *Journal of Computational and Applied Mathematics*, 53:87–108, 1994.

A. Richards and J. P. How. Robust distributed model predictive control. *International Journal of Control*, 80(9):1517 – 1531, 2007.

R. Rockafellar. *Convex Analysis.* Princeton University Press, 1997.

R. T. Rockafellar. Lagrange multipliers and optimality. *SIAM Review*, 35(2):183–238, 1993.

R. T. Rockafellar. *Network Flows and Monotropic Optimization.* Athena Scientific, Belmont, Massachusetts, 1998.

L. Scardovi, M. Arcak, and E. D. Sontag. Synchronization of interconnected systems with applications to biochemical networks: An input-output approach. *IEEE Transactions on Automatic Control*, 55(6):1367–1379, 2010.

C. Scherer and S. Weiland. Linear matrix inequalities in control. Technical report, Delft Center for Systems and Control, Delft University of Technology, The Netherlands, 2004.

I. D. Schizas, A. Ribeiro, and G. B. Giannakis. Consensus in ad hoc WSNs with noisy links - part I: Distributed estimation of deterministic signals. *IEEE Transactions on Signal Processing*, 56(1):350 – 364, 2008.

S. Srirangarajan, A. Tewfik, and Z.-Q. Luo. Distributed sensor network localization using SOCP relaxation. *IEEE Transactions on Wireless Communications*, 7:4886–4895, 2008.

G.-B. Stan and R. Sepulchre. Analysis of interconnected osciallators by dissipativity theory. *IEEE Transactions on Automatic Control*, 52(2):256–270, 2007.

I. Stojmenovic. *Handbook of Sensor Networks: Algorithms and Architectures.* Wiley, 2005.

G. Strang. *Introduction to Applied Mathematics.* Wellesly-Camebrige Press, 1986.

S. Strogatz. From kuramoto to crawford: exploring the onset of synchronization in pupulations of coupled oscillators. *Physica D: Nonlinear Phenomena*, 143(1-4):1–20, 2000.

R. Tibshirani. Regression shrinkage and selection via the lasso. *Journal of the Royal Society. Series B*, 58(1):267 – 288, 1996.

P. Trodden and A. Richards. Distributed model predictive control of linear systems with persistent distrubances. *International Journal of Control*, 83(8):1653–1663, 2010.

P. Trodden and A. Richards. Cooperative distributed MPC of linear systems with coupled constraints. *Automatica*, 49(2):479–487, 2013.

J. Tsitsiklis, D. Bertsekas, and M. Athans. Distributed asynchronous deterministic and stochastic gradient optimization algorithms. *IEEE Transactions on Automatic Control*, 31(9):803–812, 1986.

Union for the Coordination of transmission of Electricity (UCTE). System Disturbance on 4 November 2006. https://www.entsoe.eu/publications/former-associations/ucte/other-reports/, 2007.

A. J. Van der Schaft and B. M. Maschke. Port-Hamiltonian systems on graphs. *SIAM Journal of Control and Optimization*, 51:906–937, 2013.

E. Wei and A. Ozdaglar. Distributed alternating direction method of multipliers. In *Proc. of the IEEE Conf. on Decision and Control*, pages 5445–5450, 2012.

J. Wei and A. van der Schaft. Load balancing of dynamical distribution networks with flow constraints and unknown in/outflows. *Systems and Control Letters*, 2013. URL `http://arxiv.org/abs/1303.4554`. submitted March 2013.

P. Wieland. *From Static to Dynamic Couplings in Consensus and Synchronization among Identical and Non-Identical Systems.* PhD thesis, University of Stuttgart, 2010.

P. Wieland and F. Allgöwer. On synchronous steady states and internal models of diffusively coupled systems. In *Proc. of the IFAC Symposium on System, Structure and Control*, pages 1–10, 2010.

P. Wieland, R. Sepulchre, and F. Allgöwer. An internal model principle is necessary and sufficient for linear output synchronization. *Automatica*, 47(5):1068–1074, 2011.

J. C. Willems. Dissipative dynamical systems part i: General theory. *Archive for Rational Mechanics and Analysis*, 45:321–351, 1972.

F. F. Wu and C. A. Desoer. Global inverse function theorem. *IEEE Transactions On Circuit Theory*, CT-13:199–201, 1972.

W. Wu, W. Zhou, and T. Chen. Cluster synchronization of linearly coupled complex networks under pinning control. *IEEE Transactions on Circuits and Systems - I*, 56(4): 829–839, 2009.

W. Xia and M. Cao. Clustering in diffusively coupled networks. *Automatica*, 47(11): 2395–2405, 2011.

L. Xiao and S. Boyd. Optimal scaling of a gradient method for distributed resource allocation. *Journal of Optimization Theory and Applications*, 129(3):469–488, 2006.

K. Yang, Y. Wu, J. Huang, X. Wang, and S. Verdu. Distributed robust optimization for communication networks. In *Proc. of the IEEE Conference on Computer Communications*, pages 1157–1165, 2008.

R. Zamora and A. K. Srivastava. Controls for microgrids with storage: Review, challenges and research needs. *Renewable and Sustainable Energy Reviews*, 14:2009–2018, 2010.

F. Zanella, D. Varagnolo, A. Cenedese, P. Gianluigi, and L. Schenato. Newton-Raphson consensus for distributed convex optimization. In *Proc. of the IEEE Conf. on Decision and Control*, pages 5917–5922, Orlando, FL, USA, 2011.

M. Zargham, A. Ribeiro, A. Jadbabaie, and A. Ozdaglar. Accelerated dual descent for network optimization. *IEEE Transactions on Automatic Control*, 2011a. submitted (Nov. 2011).

M. Zargham, A. Ribeiro, A. Ozdaglar, and A. Jadbabaie. Accelerated dual descent for network optimization. In *Proc. of the American Control Conference*, pages 2663–2668, San Francisco, CA, USA, June 2011b.

D. Zelazo and M. Mesbahi. Edge Agreement: Graph-theoretic Performance Bounds and Passivity Analysis. *IEEE Transactions on Automatic Control*, 56(3):544–555, 2010.

Y.-B. Zhao and D. Li. Locating the least 2-norm solution of linear programs via a path-following method. *SIAM Journal on Optimization*, 12(4):893–912, 2002.

D. Zhu and P. Marcotte. New classes of generalized monotonicity. *Journal of Optimization Theory and Applications*, 87(2):457–471, 1995.